Constructing Robot Bases

Gordon McComb

McGraw-Hill
New York Chicago San Francisco Lisbon
London Madrid Mexico City Milan
New Delhi San Juan Seoul Singapore
Sydney Toronto

The McGraw-Hill Companies

Cataloging-in-Publication Data is on file with the Library of Congress.

1 2 3 4 5 6 7 8 9 0 DOC/DOC 0 9 8 7 6 5 4 3

ISBN 0-07-140852-5

The sponsoring editor for this book was Judy Bass and the production supervisor was Sherri Souffrance. It was set in Century Schoolbook by MacAllister Publishing Services, LLC.

Printed and bound by RR Donnelley.

This book is printed on recycled, acid-free paper containing a minimum of 50 percent recycled de-inked fiber.

For Jeffrey Lane McComb,
1942–2002

Big brothers never ask for the job, but mine
was perfect at it.

Photo Credits: Fig 1-4 (Parallax, Inc.); Figs 2-20, 2-21, 2-22 (Elenco Electronics, Inc.); Fig 2-23 (BK Precision Corp.); Fig 2-41 (Roto Zip Tool Corp); Figs 5-10, 5-11, 7-4 (Hemera Technologies, Inc., used under license); Figs 5-13, 5-14 (Sherline Products, Inc.); Fig 6-26, 8-11 (Du-bro Products, Inc.); Fig 8-13 (Dave Brown Products, Inc.); Figs 8-15, 8-20 (Colson Caster Corp.); Fig 8-22 (Kornylak Corp.); Fig 8-33 (The Gates Corporation); Fig 10-3 (Funhouse Productions).

CONTENTS

Contents

INTRODUCTION

A few short years ago, the typical amateur-built robot had difficulty navigating through a room. It probably cost over $1,000 to build, and it required significant engineering and construction knowledge to assemble.

Today, the landscape of robotics is vastly different.

Thanks to the collective efforts of builders in all corners of the globe, today's robots are smarter, less expensive, and easier to construct. Information is traveling faster and farther, too. New ideas spread quickly via the Internet, and a growing library of how-to magazines and books chronicle the construction and application of robots of all types.

Until now, most books on robotics could be classified into two groups. On the low end of the spectrum were the beginner's books, well suited for the roboticist in his or her formative years, but lacking the meat and potatoes for going further. On the high end were college texts containing lots of theory, but little practical building assistance.

Missing were good books in the middle . . . books for the reader who may have built a robot or two and is now looking to delve deeper.

That's how the Robot DNA series of robot construction and programming books was born. Each book concentrates on a specific segment of the robot building craft. Current books in the DNA series cover the core concepts:

- *Constructing Robot Bases* (ISBN: 0071408525) The book you're reading now is on constructing the bodies of robots using plastic, metal, and wood. It provides practical guidance on the materials and tools to use, weight distribution, fastening systems, adhesives, shop techniques, and more.
- *Building Robot Drive Trains* (ISBN: 0071408509) Motorizing with wheels, treads, and more. In this book, you'll learn how to use motors, wheels, gears, pulleys, and other mechanics to move your robots across the room.
- *Programming Robot Controllers* (ISBN: 0071408517) Giving your robot a brain is the subject of this book. Inside, you'll learn how to program robots with microcontrollers, principally the PICMicro from Micro Chip.

More books are planned in the Robot DNA series for the near future. All books are published by McGraw-Hill/Tab Electronics and written by experts.

Exploring Amateur Robotics

The field of robotics is actually fairly old, but much of the exciting technologies used in robotics are brand new. What's more, parts for making robots are far less expensive than they used to be, thanks to an increase in willing buyers. With more buyers, prices have come down.

Robotics is a huge endeavor. The Robot DNA series concentrates on a particular subfield of robotics, appropriately named *amateur robotics*. Amateur robotics is to robotics what amateur radio is to wireless communications. Both involve the same science, technologies, and even construction procedures. But like amateur radio, amateur robotics is far more contained and affordable.

So in this series, we turn the focus away from industrial and commercial robots—the kinds that build cars or travel to other planets—and toward automatons that are within the physical and financial reach of the average individual.

Amateur robotics is for anyone interested in exploring the integration of electronics, mechanics, and computers for noncommercial use. This includes hobbyists, teachers, students, backyard experimenters, and others. The DNA series is well suited to the robot enthusiast working out of his or her garage shop, using ordinary tools and materials. The books are also handy for those studying robotics in school. The various titles in the series cover all the major subsections of a robot: mechanism, electronics, and programming.

Further, these books concentrate on mobile amateur robotics—robots that are meant to move, as opposed to those that bolt to a table or floor. This distinction is a matter of scope: Discussing stationary robots would greatly increase the material the books would have to discuss. In order to provide a reasonable depth of coverage in each book, we decided to limit ourselves to those robots that rolled, walked, slithered, or otherwise moved across the floor.

Skills You Need

In this book, and others in the Robot DNA series, we make modest assumptions about your knowledge and skills. That said, we do leapfrog over the very basics, and assume you've already been exposed to those. If you're just starting out in robotics, you might want to put this book aside for a while, and start with an introductory guide. My book, *Robot Builder's Bonanza* (ISBN: 0071362967, McGraw-Hill), a perennial bestseller, is the ideal guide for introducing you to the world of robotics.

To get the most out of this and other Robot DNA series books, you'll need the following skills:

- Only a cursory understanding of schematics is required, since most robot circuits use a minimum of electronic parts. Schematic diagrams are presented for most circuits, rather than wire-to-wire drawings.
- Basic shop skills are needed for the proper operation of hand and power tools. This book does not step you through operating your Black & Decker drill, as your drill comes with a perfectly good instruction manual.
- No special attention is given to the operation of tools for electronic construction and testing. You are assumed to already know how to use them. If any tool is new to you, be sure to refer to its manual.

Tools You Need

Robotics covers many disciplines, each requiring a basketful of tools, or so it seems. Each book in the Robot DNA series requires its own basic tools. Here are the fundamental tools you need for each major discipline.

Tools for Materials Construction

When you're tired of building robots from cereal boxes, you must turn to standard construction materials: wood, plastic, and metal. These require a base set of tools if you desire a good-looking end result. These tools include the following:

- Motorized drill, with standard drill bits. If you're only interested in building robots from wood, you can make do with a hand-operated drill.
- Hacksaw. Different blades can be used to cut material.
- Screwdrivers, pliers, and wrenches, for fasteners and other hardware.

This book covers materials construction and the required tools to some degree. See Chapter 2, "Robot Tool Crib," for a more complete rundown of required and optional tools for robot construction.

Tools for Electronics Construction

No ghastly expensive tools are required for the electronics construction projects in any Robot DNA book, so you're likely to already own the ones you need:

- Soldering iron, and assorted tools, such as solder, desoldering pump, solder wick, and so on
- Volt-ohm meter, digital or analog
- Basic wiring tools: wire clippers and wire nippers

Other electronics tools are purely optional, and their use will speed you along your way to robotic nirvana. These tools include an oscilloscope, a digital waveform analyzer, a logic probe, a logic pulser, a frequency counter, and a bench-top power supply.

Tools for Programming

The Robot DNA series concentrates on robots that use microcontroller brains. Therefore, most of the required tools revolve around microcontrollers and their development systems:

- Microcontroller programmer, suitable for the chips you are using. Many programmers are designed to be used with a particular brand of microcontroller, such as the PICMicro or Atmel AVR. Other programmers are generic and can work with many programmers.
- Personal computer, with development software. Microcontrollers are programmed from a host computer, requiring a compatible PC (Windows, DOS, Linux, or Macintosh, depending on the programmer being used), software, and interface cable.

A Word (or Two) About Safety

All in all, robotics is a safe and sane hobby. Still, there's plenty of chances for you to be electrocuted, poisoned, or dismembered. So exercise care, and observe all reasonable safety precautions. We'd like you to be around to read the next books in the series!

Good safety starts with your workshop. It should be clean and uncluttered, with adequate lighting around all task areas, especially those where tools are used to cut or drill. If you work in a garage or basement, make sure the floor is dry. A wet floor can cause injury if you fall, or electrocution when using ac-operated tools.

Avoid messy and cramped work tables, where tools such as hot soldering irons can get lost or be tipped over. You'll enjoy the job of building robots much more if you avoid the added frustration of an untidy workbench.

Good ventilation keeps you awake and alert, and quite possibly, alive. A number of chemicals and other adhesives used in the construction of robots give off noxious fumes. These fumes can be both toxic and flammable. I recommend keeping a fan blowing to ensure a good exchange of air. This is particularly important when soldering. Don't breathe solder smoke.

Cutting and drilling tools are *very dangerous* if dull. Replace or sharpen as needed. Dull tools require more work, more friction, and more heat to do their job. Your grip may slip when using a dull blade or drill bit, and you could be seriously injured.

Additional tools safety tips are provided in later chapters.

Finally, wear adequate eye and ear protection. Don't even think about whacking away at some metal or plastic part without wearing safety glasses. Robot building isn't nearly half the fun with only one eye. And, if you're handling dangerous chemicals, wear gloves (as needed) and a long-sleeved shirt or lab coat to protect against skin burns.

ACKNOWLEDGMENTS

The Robot DNA series was inspired by Scott Grillo, publisher of the McGraw-Hill Professional book division. I am eternally grateful for Scott's keen vision into the world of amateur robotics, and for the opportunity to play with neat toys in the guise of writing books. Or maybe it's the other way around . . . Thanks also to Judy Bass, acquisitions editor at McGraw-Hill.

Myke Predko, book author and fellow robot enthusiast, codeveloped this book series with me. Myke likes to leave no stone unturned, yet he still manages a prolific output that a gang of writers would be hard-pressed to match.

As always, I am indebted to my friends on the Internet—most of whom I've never met—that have provided guidance, suggestions, criticisms (gulp!), ideas, and solutions. Special thanks go to Dennis Clark and Michael Owings, who also cowrote one of the books in the Robot DNA series. Special thanks again to Ed Sparks and his great 3D CAD files at FIRST CAD Library (***www.firstcad-library.com***).

Once again, my robots, and this book, would not have been even remotely possible without the continued support of my family: my wife Jennifer, my daughter Mercedes, my son Max, and my grandson Lane. I love you all.

ABOUT THE AUTHOR

Gordon McComb needs no introduction to robotics hobbyists. Mr. McComb has written more than 50 books and 1,000 magazine articles and newspaper columns, many of them on science and technology. His writings have appeared in *Popular Science*, *Omni*, *PC Magazine*, and dozens of other recognized publications. For 15 years, Mr. McComb wrote a weekly nationally syndicated newspaper column on computers and was the founder of the Robotics Workshop in *Popular Electronics* magazine. He is the author of the blockbuster *Robot Builder Bonanza*, the best-selling book on amateur robotics.

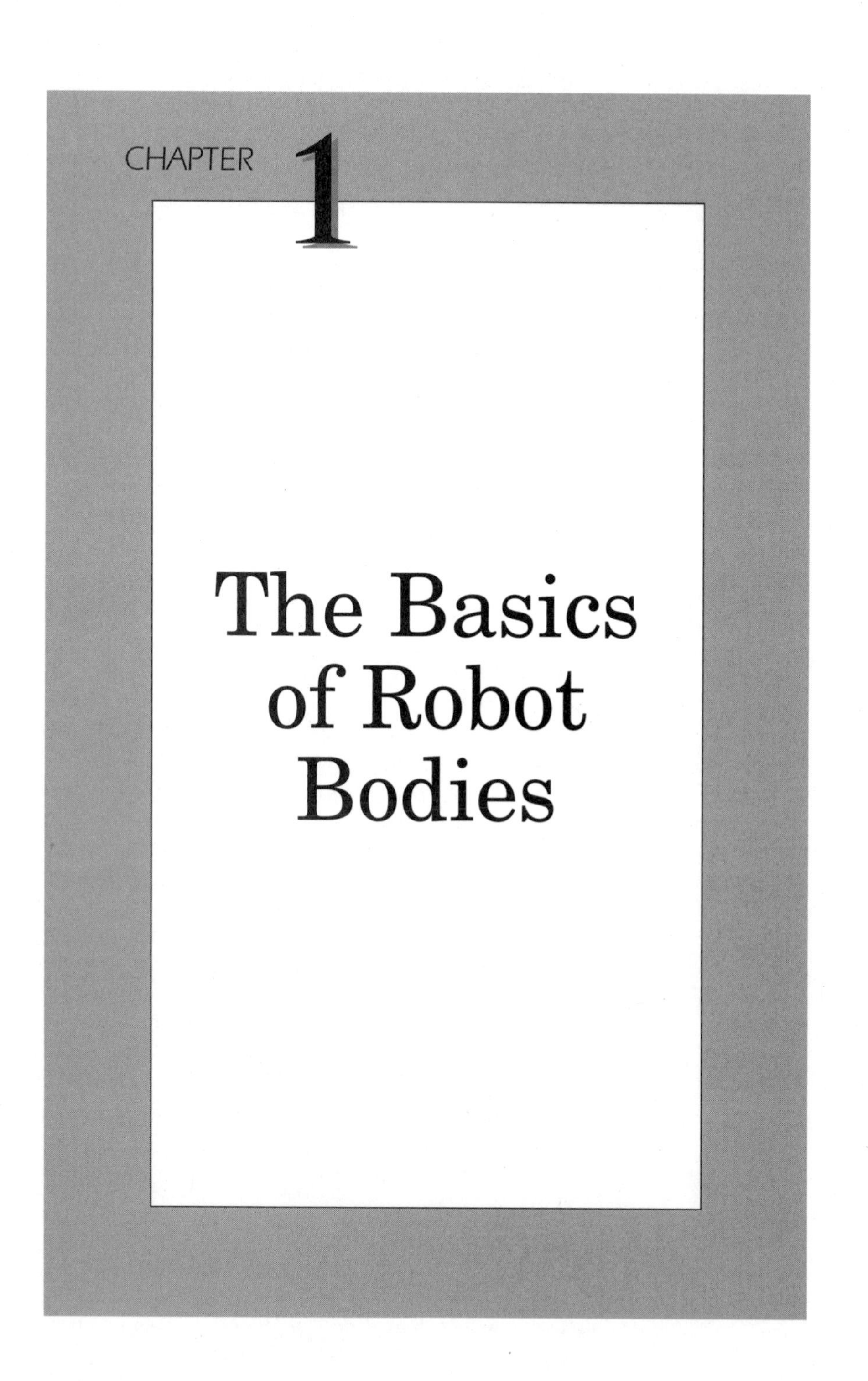

CHAPTER 1

The Basics of Robot Bodies

The robots of science fiction movies are always built using some amazing material that doesn't break and never rusts, yet seems impervious to bullets, grenades, laser beams, and green slime. Whatever this material is, the filmmakers aren't sharing their secret with the rest of us, so we're stuck using everyday stuff like plastic, plywood, and aluminum. That's okay. You can build amazing robots with these common materials, even if they can't withstand the force of an invading intergalactic armada.

In this chapter we take a look at the general choices available to us in selecting and building a robot body. The discussion begins with an overview of the methods mobile robots use to move from one place to another because this greatly influences the shape of the 'bot. The bulk of the chapter reviews various methods and materials for constructing robot bodies, including building from scratch and retrofitting old toys.

Note that this chapter introduces the most common materials used in robot construction: wood, plastic, metal, and composites. These materials—variations, choices, and how to work with them—are covered in more detail in later chapters.

Locomotion Systems

Locomotion involves the conversion of some source of energy—electricity, air pressure, steam, or nuclear power—into a mechanical action that moves a vehicle or other carriage. Consider the lowly car: Gas you put into the tank is converted to mechanical power by means of internal combustion. The gas is compressed as a vapor and explodes against a cylinder. The explosion pushes the cylinder down; this cylinder is, in turn, connected to a drive shaft, which spins the wheels. The process repeats itself thousands of times per minute.

Mobile robots use a variety of techniques to achieve motion. Most use an electric power source (usually a battery) that operates an electric motor. In the typical arrangement, like the little robot in Figure 1-1, the direction of the motors can be changed, allowing the robot to be propelled forward or backward. Other power-train techniques are used for robots, but the battery and motor pair is by far the most common, and we'll concentrate on that one for the remainder of the book.

Wheels

Wheels are the most popular method of providing robot mobility. Robot wheels can be just about any size, dictated only by the dimensions of the robot and your outlandish imagination. However, for reasons of practicality and weight, small robots usually have small wheels, less than 2" or 3" in diameter. Medium-sized robots use wheels with diameters up to 7" or 8". A few unusual designs call for bicycle wheels, which, despite their size, are lightweight but very sturdy.

Robots can have any number of wheels. Although two are the most common (see Figure 1-2), four- and six-wheel robots are also around. The robot is balanced on its wheels by one or two skids or swivel casters. Read more about wheel designs in Chapter 8, "Drive Geometries."

Figure 1-1 The vast majority of robots use electric power (usually from batteries) and electric motors for locomotion.

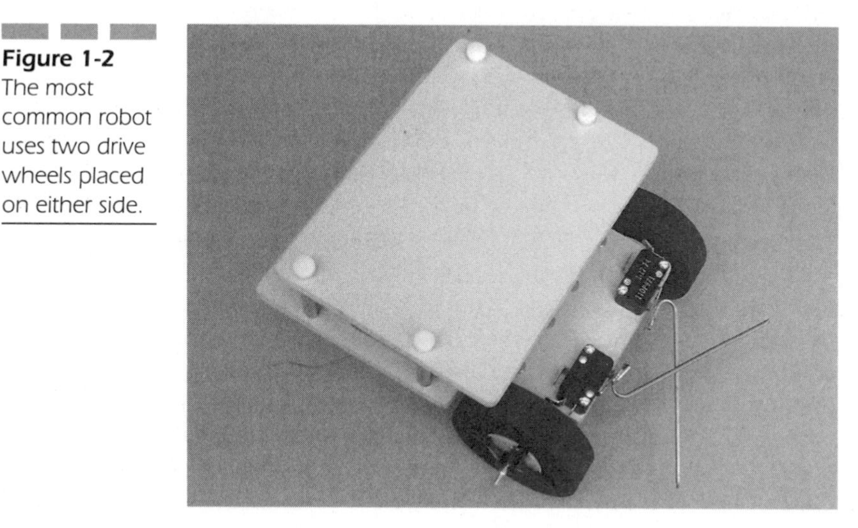

Figure 1-2 The most common robot uses two drive wheels placed on either side.

Legs

A small percentage of amateur robots are designed with legs, and such robots can be conversation pieces all of their own. Many difficulties must be overcome in designing and constructing a legged robot. First, there is the question of the number of legs and how the legs provide stability when the robot is in motion, or more critically, how to establish stability when it's motionless. Then there is the question of how the legs propel the robot forward or backward and—more difficult still—the question of how to turn the robot so it can navigate a corner.

Legged robots are a challenge to design and build, but they provide an extra level of mobility that wheeled robots cannot. Wheel-based robots may have a difficult time navigating through rough terrain, but a properly designed leg-based robot can easily walk right over small ditches and obstacles.

A few daring robot experimenters have come out with two-legged robots, but the difficulties in assuring balance and control make these designs largely impractical for most robot hobbyists. One exception is a robot that uses wide "duck" feet, and *waddles* as it walks. With this arrangement, however, turning can be inexact.

Four-legged robots (*quadrupeds*) are easier to balance, but good locomotion and steering can be difficult to achieve. I've found that robots with six legs (called *hexapods*) are able to walk at brisk speeds without falling and are more than capable of turning corners, bounding over uneven terrain, and making the neighborhood dogs and cats run for cover.

Robots with legs add many complexities. In order to devote as many pages as possible to the most critical elements of designing and building robot bodies, we'll forego a discussion of legged robots in this book. But fear not, if building walking robots is your bag, after reading this book, check out some of the other titles in the Tab/McGraw-Hill robotics catalog, such as *Insectronics: Build Your Own Walking Robot* (Williams, 2002), or my own book, *Robot Builder's Bonanza* (McComb, 2000).

Tracks

The basic design of *track-driven* robots is simple: Two tracks, one on each side of the robot, act as giant wheels. The tracks turn, and the robot lurches forward or backward. For maximum traction, each track is the same length, or somewhat shorter, than the length of the robot itself—though many variations are possible. Figure 1-3 shows a small tracked vehicle.

A track drive is practical for many reasons, including the ability to mow through all sorts of obstacles, such as rocks, ditches, and potholes. Given the right track material, traction is excellent, even on slippery surfaces such as snow, wet concrete, or a clean kitchen floor.

Figure 1-3 Tracked robots have enhanced terrain-following capabilities.

For the most part, constructing an effective track drive is harder than implementing wheels. The reason: The tracks present a large contact area. This larger contact area increases traction when moving forward or backward, but it also restricts turning. Tracked vehicles, like tanks, turn by skidding or slipping around a turning point—hence, they are referred to as having *skid-steering*. If the treads are superpliable and the surface is hard (like a kitchen floor), the added friction can greatly impair the ability of the vehicle to turn.

Comparing Locomotion Systems

You'll want to consider these pros and cons of each locomotion system as shown in Table 1-1.

Table 1-1 Locomotion systems

Application	Wheeled	Legged	Tracked
Indoors, carpet	Excellent, as long as the wheels have a radius at least double the height of the carpet nap	Good, if carpet nap is low (shag and plush can be problematic because of snagging)	Bad, the carpet fibers can get tangled in the sprockets and tracks
Indoors, hard floor or tabletop	Excellent	Excellent, the tips of the legs should be rubberized to provide at least some traction	Fair to excellent, depending on the hardness of the tracks; soft rubber tracks may not turn well on polished or hard surfaces

Table 1-1 Locomotion systems (Continued)

Application	Wheeled	Legged	Tracked
Outdoors, dirt	Fair, if ground is not too soft or uneven; foam tires pick up and hold dirt	Poor to good, depending on topology of the ground (robot may trip)	Excellent, but watch for caking up of dirt or mud in tracks and sprockets
Outdoors, grass	Poor to fair; bottom of robot must clear the height of the lawn	Fair to good, but robot should have a high-stepping gait rather than sweeping-leg gait	Excellent, although sprockets and treads can pick up grass bits
General, concrete floor	Excellent, though foam tires are sponges for dirt and dust	Excellent	Fair to good, for turnability, track material should not be too hard or too rubbery
General, asphalt floor	Fair, but surface must be fairly smooth if wheels are small in diameter	Good	Excellent

Robot Shapes, Styles, and Sizes

Robots come in all sizes. Some can fit in the palm of your hand, while others are so big it takes two people to lift them from the work table. By far, the majority are at the smaller end of the scale, weighing from 1 lb. to 15 lbs. Such robots are not only easier to build, but they are also more affordable. Their smaller size means smaller motors, smaller batteries, and smaller chassis—all of which tend to reduce price. (Of course, once size is reduced below a certain threshold, the price starts to increase again, due to the cost of miniaturization.)

Robot shapes vary, dictated mainly by the internal components that make up the machine, but also by the whims of the builder. The intended application of the robot also greatly influences its overall appearance. Limiting ourselves to the amateur robot front, most designs fall into one of the following categories:

- Turtle (also called desktop)
- Rover
- Walking
- Arms and grippers (also called appendages)
- Android and humanoid

Turtle or Desktop

Turtle or so-called desktop robots are simple and compact. Like the wording implies, these creations are designed primarily for "tabletop robotics." Turtlebots get their name because their body somewhat resembles the shell of a turtle. Researcher F. Grey Walter used the term to describe a series of small robots he envisioned and built in about 1948. In popular usage, turtle robots also borrow their name from a once-popular programming language, Logo turtle graphics, adapted for robotics use in the 1970s.

The turtle category, which represents perhaps 75+ percent of amateur robots, is popular among those involved in noncombat competitions—such as maze following or robotic soccer. Turtle robots are most commonly powered by a rudimentary brain (*BEAM—Biology, Electronics, Aesthetics, Mechanics*—robots, which are based on simplified electronics, are a good example), or by a small single-chip computer or microcontroller.

Rover

The *rover* category includes any of a group of rolling or tracked robots designed for applications that require some horsepower, such as vacuuming the floor, fetching a can of beer or soda, or mowing the lawn. These robots are too big to play with on a desktop. Sizes range from that of a small step stool to desk size. The death-match combat robots popular on TV are typically robots in the larger end of the rover spectrum, where weight is important to winning.

Because of their larger size, rover robots can be powered—brainwise—by everything from a simple transistor to a desktop computer. Old laptops, particularly the monochrome models, are popular among many robot builders, because they can run DOS or early versions of Microsoft Windows, and they can be interfaced to the robot via standard parallel and serial ports. Old PC motherboards can also be pressed into service for running the rover, though this approach is falling out of favor because of the relative abundance of older, used laptops on the market.

Walking

A *walking* robot uses legs, not wheels or treads, to move about. Most walker 'bots have six legs, like an insect, as the six legs provide excellent support and balance. However, gaining in popularity are two- and four-legged walkers, both for scratch-build projects and commercial kits. An example of a two-legged walking robot kit is the Toddler, from Parallax, Inc. (Figure 1-4), makers of the ever-popular Basic Stamp microcontroller. Despite walking on just two legs, the Toddler exhibits excellent balance.

Walking robots require a great precision in building. The design of the typical robot that rolls on wheels or even treads allows for a certain amount of slop. Devices that run on wheels are inherently simpler than the cams, links, levers, and other mechanisms used for walking. For this reason, beginners in the robot-building trade should opt for wheeled designs first to gain experience, even if the walking robot looks cooler.

Note that two-legged walking robots that resemble people are classified in their own category, which considers the technological difficulties in designing and building them. Constructing a small two-legged robot that hobbles along the desk is one thing; creating a C-3PO-like robot is quite another, even if your name is Anakin Skywalker.

Figure 1-4 The Toddler two-legged walking robot from Parallax, Inc. Photo courtesy Parallax, Inc.

Arms and Grippers

Arms and *grippers* are used by themselves in stationary robots or can be attached to a mobile robot. Arms can be considered any appendage of the robot that can be individually and specifically manipulated; grippers (also called *end-effectors*) are the hands and fingers and can be either attached directly to the robot or to an arm.

The human arm has joints that provide various *degrees of freedom* (DOF) for orienting the appendage in most any direction. Likewise, robotic arms have DOF. In most designs the number of DOFs is fairly limited (between one and three). In addition to DOF, robot arms are further classified by the shape of the area that the end of the arm (where the gripper is) can reach. This accessible area is called the *work envelope*. Classifications include the following:

- *Revolute coordinate* arms are modeled after the human arm, so they have many of the same capabilities. This type is a favorite design choice for hobby robots.
- The work envelope of the *polar coordinate* arm is half-sphere shaped. The arm is basically a tilting lever built on a turntable and is quite flexible in being able to grasp a variety of objects scattered around the robot.
- The *cylindrical coordinate* arm looks a little like a robotic forklift. The forearm is attached to an elevator-like lift mechanism, which moves up and down this column to grasp objects at various heights.
- The work envelope of a *Cartesian coordinate* arm resembles a box. It is the most unlike the human arm and it is unique in that it has no rotating parts. The base consists of a conveyer-belt-like track, which moves the elevator column.

Android and Humanoid

Android and *humanoid* robots are specifically modeled after the human form: a head, torso, two legs, and possibly one or two arms. In current usage, the terms android and humanoid are not the same: An *android* is a robot designed to look as much like a human being as possible, including ears, hair, and even an articulated mouth. A humanoid robot is one that shares the basic architecture of a human—bipedal (two legs), head at the top, two arms at the side—but is not meant to be a physiological replica.

Whether a true android or a humanoid, this category of robots is something of a Holy Grail for robot builders, and is also the most difficult to accomplish. Few amateur builders attempt, and still fewer succeed, in building an autonomous, walking, human-like robot. Of the humanoid robots that are built, many are stationary: They're a torso and head and possibly an arm.

The problems facing the designer and builder of the android or humanoid robot are boundless, and cost of construction is *much* higher than other forms or robots. That said, given time, a hobby-level humanoid robot that walks is surely on the horizon. Though frightfully expensive, such robots are already being demonstrated.

The compelling rationale of human-shaped robots is that because the machine walks on two legs, it can live and work in the same environment as a human. Contrast this with a robot that must roll on wheels or tracks: Stairs become difficult, and even clothing discarded on the floor can impede the motion of the robot.

Materials for Robot Building: An Introduction

With locomotion system and body type and size firmly in mind, it's time to turn to both the method of construction and the materials used for building robots. It may be a surprise to some, but building a robot is a tad more complicated than going out to the garage and cutting up a hunk of pine. There are other—and frankly better—methods.

To begin the discussion, you have a choice of building the robot *from scratch*, using raw materials like plywood or sheet metal, or, if you prefer, you can *adapt* some ready-made product to serve as the base of your robot. Inexpensive housewares, hardware items, and toys can be used in various creative ways to make robot building faster and more economical.

Scratch-Build Using Metal, Plastic, Wood, and Composites

When someone cooks from scratch, it means to prepare food using basic ingredients such as flour, milk, butter, baking powder, salt, and garlic—in this case, garlic-flavored pancakes. Similarly, in robot construction, building from scratch means using raw materials such as plastic, wood, and metal to concoct a robot body. More often than not, the body of the robot is cut from a larger piece of material and shaped into the desired form. Round, oval, and square robot bodies are common.

Do note that I make a distinction between scratch building and adapt building. The latter (my own made-up phrase) means taking commercially available products and adapting them to construct a robot. Because toys can be a gold mine in adapt building, this subject is separately discussed at much greater length later in this chapter.

NOTE: *Building from scratch assumes the availability of at least a minimum set of tools, including a saw, drill, and screwdriver, and, of course, a place to use them. As such, scratch building is best suited for those robot builders who have these tools available and a workshop to ply their craft. Those with limited tools and little workshop area should consider building their robots from adapted parts or toys. This may seem obvious, but keeping this recommendation in mind helps match the robot-construction method with the available resources.*

Choosing the Material

Raw construction materials should be chosen for their suitability for the job, as well as their machining and shaping requirements, *not* by their price or availability. Bear in mind the tools and skills you have, and match the materials to them. A cheap piece of old wood is hardly a good choice for a robot body, though it may be free and no farther away than the shed. The wood may be warped and weak or full of termites; if your time and temperance matter to you, avoid materials that cause extra work and frustration.

Wood Wood is reasonably inexpensive and can be worked with using ordinary tools. Avoid soft plank woods, such as pine and fir, because they are bulky for their weight. Instead, go with a hardwood such as ash or birch, but stay away from the very dense hardwoods. Oak is too heavy, and its density makes it a chore to cut and drill.

I prefer hardwood plywoods designed for model building. Though more expensive—about \$5 to \$8 for a 12" × 12" square—the plies are sturdy and won't delaminate. This makes the wood resistant to cracking and flaking. Thickness ranges from 1/8" to over 1/2", with 1/4" being a good compromise for typical applications. Hardwood plywood is available at better-stocked hobby stores, but most folks I know purchase it by mail order from specialty retailers. These include woodturning supply, model shipbuilding, and taxidermy catalogs. Examples include Constantine's Wood Center at ***www.constantines.com*** and Penn State Industries at ***www.pennstateind.com***.

For a unique look, consider using a so-called exotic wood meant for millwork, which is available from woodturning supply catalogs. These include bubinga, teak, walnut, and rosewood. These are hardwoods designed for turning on a lathe or for carving or milling. They tend to be quite expensive: $8 for a 1" x 1" x 36" length. Still, they are well suited for such roles as sturdy risers that separate the decks of a multiple-tier robot. And, when properly sanded and finished, they have the look of fine furniture, adding uniqueness to your robot.

Not all hardwoods are "hard"—the term actually applies to the wood from a deciduous tree (which loses its leaves every season). Softwoods come from non-deciduous, or evergreen, trees. With this in mind, cottonwood (*Populus fremontii*) and balsa (*Ochroma pyramidale*) are both hardwoods because they are from deciduous trees. Yet they have low densities, so they are somewhat soft and easy to work with. Balsa is commonly used to make model airplanes because it's strong and lightweight. It can be useful in robotics as a way to create quick engineering samples as well as struts for structural strengthening.

Plastic Plastic is the material of choice by manufacturers because it can be readily molded to shape. The typical plastic-molding process involves molten plastic injected under high pressure into specially made metal forms. Injection molding is a manufacturing technique that is not readily adaptable by the amateur robot builder. Instead, our plastic material of choice is the raw shape: sheet, bar, rod, and so forth, which is then cut into the desired form. These sheets, bars, rods, and more can be purchased at home improvement stores, specialty plastics retailers, and sign makers.

Literally thousands of plastics can be found on the market, and each is designed to be best suited for a particular application. For robotics, just a small handful of plastic materials that are both affordable and readily available. These same plastics also tend to be the ones most easily worked when using standard shop tools:

- *Acrylic* is used primarily for decorative or functional applications, such as picture frames or salad bowls. It can also be used in low-stress structural applications, such as robot bodies, as long as limitations for weight and impact shock are observed.
- *Polycarbonate* is similar in looks to acrylic, but is considerably stronger. This plastic is a common substitute for window glass; because of its increased density, it's much harder to work with.
- *Polyvinyl chloride (PVC)* is familiar to anyone who has installed a water irrigation system. Though white pipe is a common form of PVC, it's just one of many. PVC plastic is routinely available in all shapes, including colored sheets of various thicknesses. PVC sheets are often made using a gas expansion process that "bulks up" the plastic, making it lighter. These

expanded yet rigid sheets are used to cut out shapes for making signs. Expanded rigid PVC (see Figure 1-5) can be cut, drilled, and even sanded like wood—in fact, this material is often used as a substitute for wood trim.

- *Urethane resin* is a common component in casting plastics, such as that used with fiberglass. Although you can buy already cured bars, rods, and other shapes of urethane, you will more likely use the liquid resin to mold your own shapes.
- *Acetal resin* is referred to as an engineering plastic. A typical use is as parts that are turned or milled and do not require the strength of steel or aluminum. Acetal resin (sometimes referred to as *Delrin*, a popular trade name for an acetal resin product) is softer than metal so the work goes faster. Yet the plastic is surprisingly dense and strong. If you have a small lathe, you can use acetal resin to make components for your robot. Even if you don't have a lathe, you can cut the raw acetal resin—it comes in rods, bars, and sheets—into useful shapes.

Acrylic, polycarbonate, PVC pipe, and liquid urethane resin are commonly available at home improvement stores. Expanded rigid PVC sheets and acetal resin can be purchased from a plastics specialty retailer. Check the Yellow Pages under the various *Plastics* headings.

Metal Metal is the archetypal material of a robot, though it's not always the ideal choice. Metal is among the most expensive materials for robots—in terms of both cost and weight—and is harder to work unless you have the proper tools and skills. That said, metal is a must if your robot will be bashing other robots in a death-match contest. Metal is ideal for rugged outdoor use or any of a number of other valid reasons that call for a strong body in a small package.

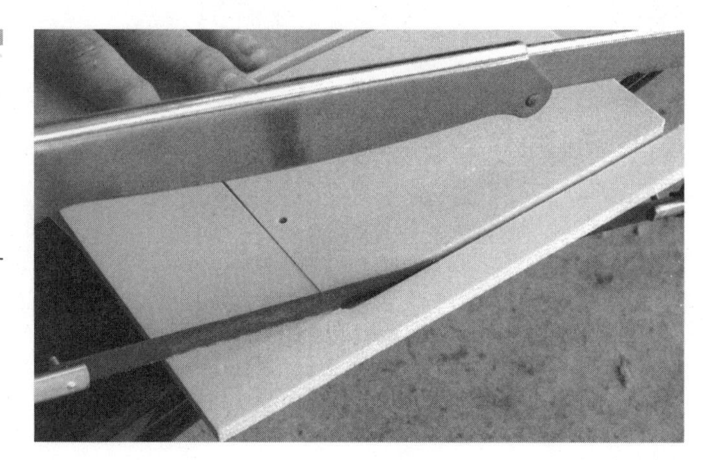

Figure 1-5 Rigid expanded PVC is an ideal material for constructing robot bases.

For robots, aluminum and steel are the most common metals. Aluminum is a softer metal and is therefore easier to work with, but steel is several times stronger. In any case, because of the inherent strength of metal, robot bodies can be made using sheets, bars, rods, channels, and other shapes. You can take one of two general approaches to metal construction in robots:

- A *frame* provides the base of the robot and lends its support. The frame can be flat or box shaped. A flat frame has four corners and four sides and provides a convenient platform for motors and other components. A box-shaped frame is just what its name implies: a 3-D box with six faces. The box frame, like that in Figure 1-6, is particularly well suited for larger robots or those that require extra support for heavy components.
- A *shaped base* is a piece of metal cut out in the shape of the robot (see Figure 1-7). The metal must be rigid enough to support the weight of the motors, batteries, and other parts without undue bending or flexing. For small tabletop robots, a thickness of 22 or 24 gauge steel, or 1/32" aluminum, is usually sufficient. The weight of the robot increases dramatically with thicker sheets. Some very capable robots are basically a piece of sheet metal on wheels, with a laptop PC resting on top.

Composites Composites, as used in this book, take three primary forms:

- Any laminated material, typically in sheet form, that combines wood, paper, plastic, or metal, relying on the intrinsic properties of each in order to

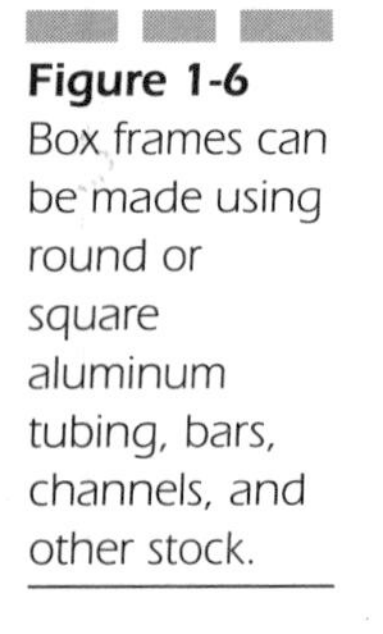

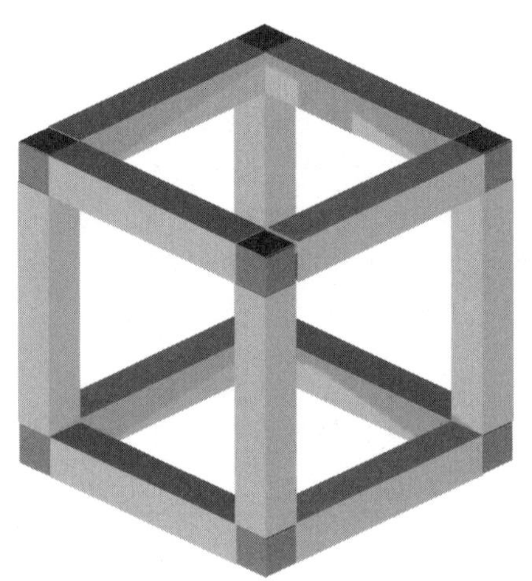

Figure 1-6 Box frames can be made using round or square aluminum tubing, bars, channels, and other stock.

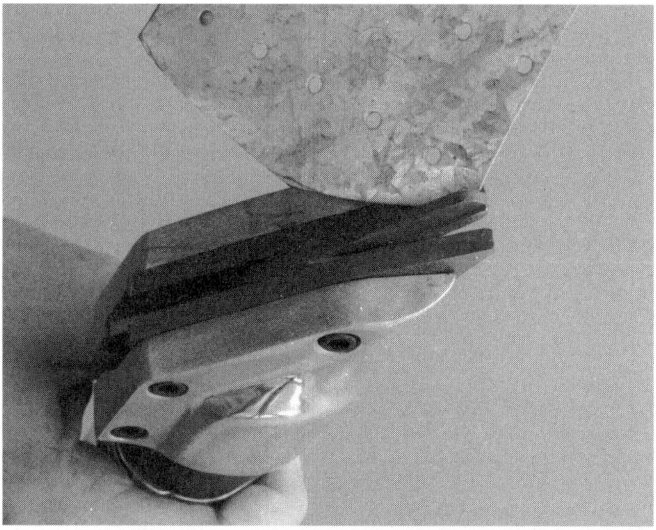

Figure 1-7 Round or oblong bases can be cut from a steel or aluminum sheet.

increase rigidity or strength. A common laminate composite is foam board (such as Foamcore brand), which is made by sandwiching a core of springy foam between two pieces of stiff paper. Other laminates might combine wood and metal, plastic and paper, or most any other combination of materials.

- Any material using fiberglass and a resin. Sometimes a filler of metal, fabric, or carbon is added to the resin to give it extra strength.
- Any material using carbon or graphite for strength (these materials may or may not contain other components). A good example is carbon composite tent poles that are lightweight, flexible, and incredibly strong.

Why use a composite? Because of its weight-to-strength ratio. The strength of composites is relative—some composites, such as foam board, are lightweight but not very strong, and are best used for creating mockups or for reinforcing wood or plastic parts. You can cut it with just a knife and a straightedge. Other composites are both lightweight and very strong (carbon composites), but require specialty tools for cutting and drilling.

Two disadvantages, at least of the stronger composites, are cost and availability. Most composite materials are only available from specialty retailers and industrial suppliers.

Adapt Build Using Manufactured Parts

Adapt build is my phrase for constructing robot bodies using various items meant for completely different purposes. Toys are a common commercial product for retrofitting (and are discussed separately later), but you can use plenty of other everyday objects for robot building. Examples are

- **Plastic fishing tackle box** Wheels are mounted to the sides of the box; motors, batteries, and other critical components are placed inside. The box can be popped open to gain internal access.
- **Plastic storage containers** Available in square, round, and other shapes, these durable plastic boxes—available in the housewares section of any department store—can be used with or without their press-on lids. Plastic boxes are available from small snack size to big shoebox size.
- **Small dorm-size trashcans** Just large enough to hold a Big Gulp, these trashcans have a convenient cylindrical shape and removable top. Great for building miniature R2-D2 'bots.
- **Computer mice** A discarded computer mouse makes a great body for a microminiature robot. Almost all mice can be disassembled by removing one or two screws on the bottom. After removing the circuit board, mouse ball, cable, and switches, you can install small motors, a small battery, and a one-chip brain.
- **Compact disc** Save the world's landfill and use these 4.7" diameter discs for robot bases. Use care when drilling holes in the plastic; the material can shatter into very sharp pieces.
- **Solderless breadboard** Solderless breadboards are used to experiment with circuits before using more permanent solder and wire-wrap construction. Mount motors and wheels on the underside of your solderless breadboard, and you can create a versatile and ever-changeable mobile robot.
- **Plastic project boxes** These boxes, sold by RadioShack and other electronics stores, are made to hold custom electronics projects. The boxes come with removable metal or plastic lids to allow access to the inside. The plastic is easily drilled for mounting motors and other parts.
- **Sports safety (bicycle, skate) helmet** I first saw this in a prototype robot made by British electronics guru Tony Ellis. Safety helmets, in both child and adult sizes, are typically made of fiberglass, and with a dense compressed foam liner. Select a helmet with a shape that lends itself as a shell for a turtle-size 'bot.

- **Clear or colored display dome** Also called a hemisphere or half-round dome, display domes can be purchased in sizes from about 2" to over 12" in diameter. The dome can be used as the body of the robot or as a cover to protect its electronics. A robotic "ball" can be made by gluing two domes together. The wheels of the robot spin the ball, which in turn rolls on the floor.
- **Metal hardware parts** These include T-braces used for lumber framing in houses. Sizes and shapes vary greatly; take a stroll down the aisles at the hardware store and you're sure to find plenty of candidates.
- **Halloween costumes** Let your imagination soar. Wait until November 1 to save money on unsold costumes.
- **PCV irrigation pipe** All forms of polygonal frames can be constructed using PVC irrigation pipe. Most hardware and plumbing supply stores carry PVC pipe in various sizes and wall thicknesses. Select the pipe based on the size and weight of the robot. Obviously, you'll need larger and thicker pipe for big and heavy robots.

Using Toy Parts for Building Robot Bases

Toys can be a cheap source of useful components, from wheels and gears to motors and blinking lights. Most toys are mass produced, so they're relatively cheap for what you get. And because the popularity of toys tends to run in cycles, you can often find great deals on discontinued items.

Though there are literally hundreds of types of toys, we'll concentrate on just three categories:

- Construction toys
- Ready-made toys that can be disassembled for their parts
- So-called dollar-store toys, which are a good source for basic parts

Construction Toys

Construction toys are those that come as a box of parts. Most are the snap-together type.

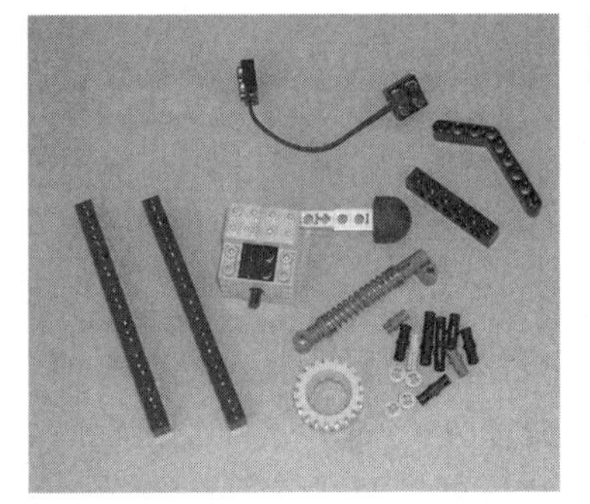

Figure 1-8

LEGO

LEGO has become the premier construction toy for both children and adults. The LEGO Company, parent company of the LEGO brand, has expanded its line to educational resources, making the ubiquitous LEGO bricks common around the world. LEGO also makes the Mindstorms, a series of sophisticated computerized robots.

The LEGO Technic line provides additional parts for robot construction. The pieces are engineered to fit the classic LEGO bricks but add useful beams, connectors, and other parts to provide greater flexibility. A LEGO Technic beam has holes drilled through its sides for attaching to connector pieces. These holes are perfect for mounting LEGO components onto the rest of your robot.

Figure 1-9

MEGA BLOKS

The MEGA BLOKS toys span a variety of play sets, from their Dragon fantasy line to their motorized robots. The MEGA BLOKS toys are competitively priced, and though the company is careful about touting it, the construction pieces are more or less LEGO compatible.

Certainly, one use of MEGA BLOKS is as a low-cost alternative for some basic LEGO pieces, but for the robot builder, you'll be interested in such products as their Battle Bloks RC, which are radio-controlled, motorized platforms that can be readily converted to robotics use. Battle Bloks use a six-wheel, all-terrain design along with dual motors that can be used to create a powerful and zippy robot. The Battle Bloks bases are readily hackable by adding your own control electronics.

If you've played with LEGO for any length of time, especially LEGO Technic, then you know certain building techniques allow you to create self-locking constructions that defy coming apart accidentally. MEGA BLOKS sets lack many of the Technic-style pieces, and its parts are basically the core LEGO bricks and plates. If you want a more permanent creation, you may need to glue the pieces together. Whereas LEGO uses ABS plastic for its pieces, MEGA BLOKS uses polystyrene, the same material found in plastic model kits. For a strong bond, you can use plastic model cement; for a less permanent bond, a flexible silicone-based or silicone-like adhesive, such as Household Goop, can be used.

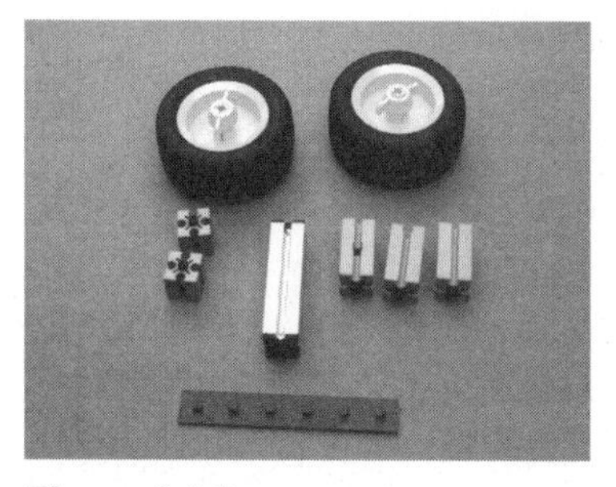

Figure 1-10

Fischertechnik

The Fischertechnik kits, made in Germany, are favored in higher-ed classes, and even some colleges and universities. More than toys, Fischertechnik kits offer a snap-together approach to making working electromagnetic, hydraulic, pneumatic, static, and robotic mechanisms.

All the Fischertechnik parts are interchangeable and attach to a common plastic base plate. You can extend the lengths of the base plate to just about any size you want, and the base plate can serve as the foundation for your robot. You can use the motors supplied with the kits or use your own motors with the parts provided.

Figure 1-11

K'NEX

K'NEX uses unusual half-round plastic spokes and connector rods to build everything from bridges to Ferris wheels to robots. You can build a robot with just K'NEX parts or use the parts in a larger, mixed-component robot. For example, the base of a walking robot may be made from a thin sheet of aluminum, but the legs might be constructed from various K'NEX pieces.

A number of K'NEX kits are available, from simple starter sets to rather massive special-purpose collections (many of which are designed to build robots, dinosaurs, or dinosaur robots). Several of the kits come with small gear motors so you can motorize your creation. The motors are also available separately.

Figure 1-12

Erector Set

Erector Sets, now sold by Brio, were once made of all metal. Now they commonly contain both metal and plastic pieces, in various sizes, and are generally designed to build specific vehicles or other projects. Useful components of the kits include prepunched girders, plastic and metal plates, tires, wheels, shafts, and plastic mounting panels. You can use any as you see fit, assembling your robots with the hardware supplied with the kit, or with 6-32 or 8-32 nuts and screws.

Over the years the Erector Set brand has gone through many owners. Parts from old Erector Sets may not fit well with new parts, including, but not limited to, differences in the threads used for the nuts and screws. If you have a very old Erector Set (such as the ones made and sold by Gilbert), you're probably better off keeping it as a collector's item rather than raiding the set for robotic parts. Similarly, today's Meccano and Brio sets are only passably compatible with the English-made Meccano sets sold decades ago. Hole spacing and sizes have varied over the years, and mixing and matching is not practical, or desirable.

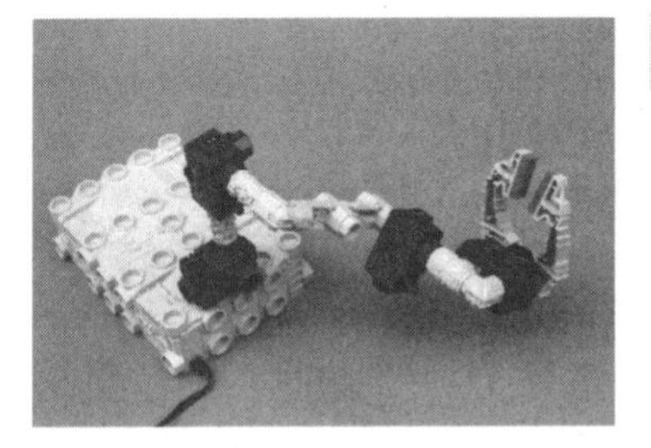

Figure 1-13

Robotix

The Robotix kits, originally manufactured by Milton-Bradley and now sold by Learning Curve, are specially designed to make snap-together walking and rolling robots. Various kits are available, and many of them include at least one motor; you can buy additional motors if you'd like. You control the motors using a central switch pad.

The structural components in the Robotix kits are molded, high-impact plastic. You can connect pieces together to form just about anything. You can cement the pieces together to provide a permanent construction.

Figure 1-14

Capsela

Capsela is a popular snap-together, motorized, parts kit that uses unusual tube and sphere shapes. Capsela kits come in different sizes and have one or more gear motors that can be attached to various components. The kits contain unique parts that other put-together toys don't have, such as plastic chain and chain sprockets and gears. Advanced kits come with remote-control and computer circuits. All the parts from the various kits are interchangeable.

The links of the chain snap apart, so you can make any length chain you want. Combine the links from many kits and you can make an impressive drive system for an experimental lightweight robot.

Figure 1-15

Inventa

UK-based Valiant Technologies offers the Inventa system, a reasonably priced construction system aimed at the educational market. Inventa is a good source for gears, tracks, wheels, axles, and many other mechanical parts. Beams used for construction are semiflexible and can be cut to size. Angles and brackets allow the beams to be connected in a variety of ways. It is not uncommon—and in fact it's encouraged—to find Inventa creations intermixed with other building materials, including balsa wood, LEGO pieces, you name it. Inventa products are available from distributors, which are listed on the Valiant Web site at ***www.valiant-technology.com***.

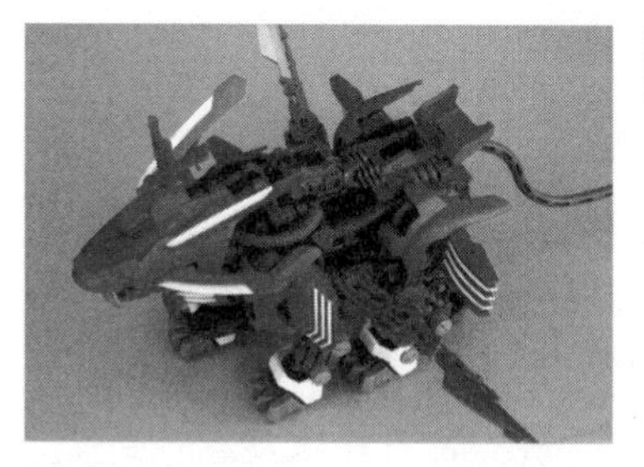

Figure 1-16

Zoids

Tomy's line of Zoids are construction toys designed to build a specific mechanical creature. Many are motorized, either with a battery-operated motor or with a wind-up mechanism. Zoids toys use only snap-together assembly.

Although each Zoids has a specific design, many of the parts are common to each kit, and you can use them to build your own creation. You can also use the electric or wind-up motors for your own robots.

Additional Construction Toys to Consider There are numerous other construction toys that you may find handy. Try the nearest well-stocked toy store or a toy retailer on the Internet for the following:

- Expandagon Construction System (Hoberman).
- Fiddlestix Gearworks (Toys-N-Things).
- Constructo Botix (Wow-Wee).
- PowerRings (Fun Source).
- Zome System (Zome System).
- Construx (no longer made, but sets may still be available for sale).
- Fastech construction sets (no longer made) were among the best parts assortment you could buy.

Scavenging Ready-Made Toys

Ready-made toys are those that come pre-assembled but can be taken apart for useful parts. Good examples are Bio-Bugs (Wow Wee), Furby (Tiger Electronics), and various motorized Tonka (and similar) tractors and trucks.

These and other toys like them can often be found on the clearance aisles, at garage sales, and even in thrift stores. At full price, many of these toys don't have enough useful parts in them to justify their cost, so always strive to purchase them used or at a discount.

Best Traits of Scavenged Toys Here are the common elements of the ideal ready-made toy for scavenging:

- *It is motorized*, preferably with an electric motor. This category includes small *four-wheel-drive* (4WD) cars and trucks: They contain one motor that drives all four wheels at a time. You can readily adapt two of these toys for use in a traditional two-motor robot design. Just pop off the wheels on one side of the motor and mount the two motor and wheel assemblies onto your robot. Each motor and wheel assembly can be independently operated under electronic control.
- *It uses a self-contained, speed-reduction gearbox* rather than gears mounted in the chassis of the toy. The latter is actually far more common, and not as useful, because you cannot readily repurpose the gearbox in your own creations. Self-contained gearboxes can be yanked out of the toy and implanted directly into your robots.
- *If motorized, it uses electronic* rather than mechanical *control for the motors*. With some of the Tonka and other tractor-type toys, the motors are controlled by operating levers, which are really mechanical links. To save costs, these toys use a single motor that is mechanically coupled to various wheels, treads, joints, and other articulations. You likely won't be able to adapt these often-elaborate mechanical links in your robot without considerable effort, so it's best to just concentrate on the parts you can readily pull out and reuse.
- *It uses screws for assembly.* This enables you to disassemble the toy by removing the screws. Toys that are made by gluing, hot-bonding (a kind of spotwelding for plastic), or using hydraulic press-on joints are harder to take apart.

Hacking Motorized Vehicles Although many toys simply beg to be yanked apart for their guts, others are useful in much of their original form as robot bases. Motorized *radio control* (RC) cars and treaded vehicles (Tonka or New

Bright) are among the most common ready-made toys that are used whole as a robotic platform. The best such toys use a separate body and chassis; you can remove the body and use only the bare chassis. It's easier to mount your robot parts to the chassis, and of course your robot looks less like a 1975 Le Mans (dark blue with white racing stripe) or yellow Caterpillar earthmover.

Literally hundreds of motorized RC cars and toy treaded vehicles are on the market, but not all toys are universally available to everyone, or would be considered inexpensive enough by all to hack. So, if you're interested in using a motorized car or other vehicle as a robot base, know it's a project that is best suited for those who enjoy experimentation and tinkering. Because the toy may be ruined in your hacking efforts (it happens to even the most seasoned robot builder), it's best to use only toys purchased at a discount, either on clearance, at a thrift store or garage sale, or even pulled from your child's closet of forgotten stuff.

For the hopelessly curious, I present several inside shots (Figures 1-17 through 1-19) of a typical treaded vehicle, to show the mechanical construction. These vehicles tend to move too slowly for my taste, so they're not my favorites to hack. But they may be just the ticket for your next project.

Hacking Versus Adapting In this case, hacking is used in the good sense—taking something, disassembling it, and turning it into something else. You've already seen how ready-made toys like treaded vehicles or RC cars can be hacked apart to serve as robot bases or general parts for your robots.

You don't need to tear everything apart to use it. Sometimes a quick-and-easy adaptation is all that is required. You need not disassemble the toy to make use of it. Case in point is the wireless remote-controlled sets from Rokenbok. These

Figure 1-17 A treaded toy construction vehicle, as it came from the store

Figure 1-18 A close-up of the motor system of the toy tracked vehicle

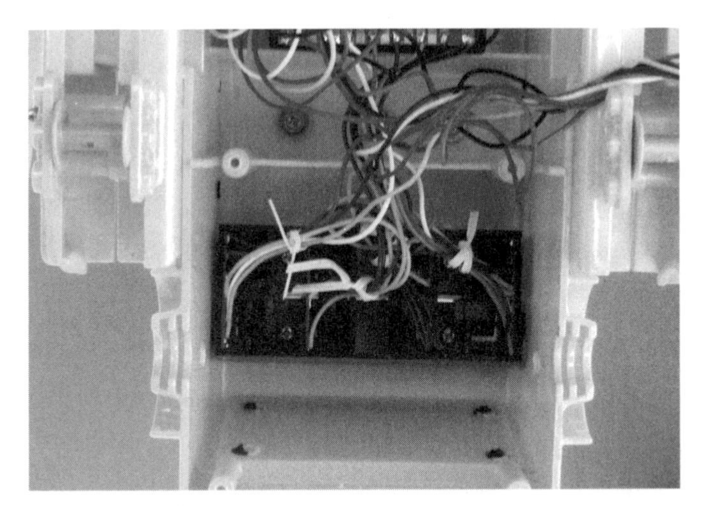

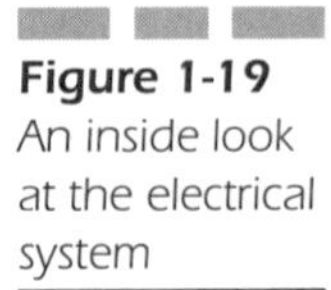

Figure 1-19 An inside look at the electrical system

small vehicles are controlled via a centralized radio tower—you can operate up to four vehicles at the same time. Each vehicle is equipped with a receiver that operates at one of eight specific frequencies. You select the receiver and therefore which vehicle you want to operate on the handheld controller.

The Rokenbok vehicles and other parts aren't cheap; each vehicle costs from $50 to $90. At these prices, you probably don't want to hack your Rokenbok toys for their motors, gears, and wheels. You can buy such parts for less. Rather, consider adapting the complete Rokenbok vehicles as robot bases. The body of most vehicles can be removed using a screwdriver. You can then mount your own robot body to the chassis.

Information about the internal architecture of the Rokenbok system is hard to find. With some work, the vehicles could be adapted to computerized control with any of the following:

- Hack the "Smart Port" on the Rokenbok radio tower (they call it the Control Center). The Smart Port, found only on earlier Rokenbok radio towers, was never documented.
- Communicate with the vehicles via the same radio frequencies and data format the Control Center uses. It's a rather easy task to determine the operating frequency of the Control Center if you have access to an RF-spectrum analyzer.
- Discard the control electronics in the Rokenbok vehicles and connect a microcontroller and an H-bridge directly to the motors. If you perform this surgery carefully, you could insert a switch or jumper block to allow you to switch between regular RC mode and direct mode. That way your kids won't yell at you when they've discovered you've hacked yet another one of their toys.

Note that while the name Rokenbok conjures images of some European toy company, the toys are actually made in the United States. At this writing, they are available at such retailers as Toys "Я" Us, Target, and numerous sites online. See ***www.rokenbok.com*** for more details.

Scavenging Dollar-Store Toys

A relatively new kind of retail store in North America and elsewhere is the dollar store, so named because everything costs a dollar, or 99¢, or something close to it. More than likely, there's at least one dollar store in your neighborhood, and this store is practically guaranteed to sell a wide assortment of really cheap toys. And I do mean cheap, in both cost and manufacture. These toys are made for pennies, and imported in very high quantities. Although they're not well-made toys, some contain parts that can be used in your robots.

That's the good news. Even better news is that each dollar store (and dollar-store chain, such as 99 Cents Only) carries its own line of imported goods. So, even if one store doesn't carry what you need, the next one may. However, it also means that the pickins are sporadic; what I can find at my local dollar stores may be different from what you can find. It's all one big crapshoot.

As you scan the wall of toys at the store, be on the lookout for the following:

- **Electric motorized vehicles** Some are self-running, but others are operated by a wired remote. I've found a significantly high failure rate in these toys, usually from some problem in the remote (which is also an AA battery holder). You may not be able to use the remote, but you can yank apart the toy for its motor, gears (if any), wheels, axle, and other parts.

- **Wind-up motorized vehicles** You can make your own robotic wind-up toys with the wind-up mechanism for a car or truck. Look for toys with a self-contained winder (most have this). I prefer the mechanisms that are wound by turning a knob, but there are also pull-back (friction) and pull-string assemblies.
- **Miniature nonmotorized vehicles** Rob them of their wheels, which can be used as small support casters or as drive wheels for miniature robots.
- **Miniature skateboard or scooter toys** These also have nice wheels you can use on smaller robots.
- **Big plastic toy vehicles** These have larger 2" to 4" diameter wheels. The better wheels use a softer rubber, but the hard plastic wheels can also be useful. Be sure the wheels can be pulled off the axle without damage.
- **Friction spark guns, vehicles, and other toys** For decoration purposes, remove the spark wheel and mount it on your robot for a rad design. You can spin the wheel with a small motor. Remember that sparks are white; if you want colored sparks, mount some colored plastic gels over the wheel.

The vast majority of dollar store toys use a screw assembly. You need a set of size 0 and 00 (jeweler's) Phillips screwdrivers and possibly a pair of needle-nose pliers and nippy cutters. I regularly save the tiny screws (they're size 0 to 000); they're self-tapping and come in handy for various lightweight fastener jobs—plus they're effectively free.

Occasionally, you'll run into toys that use rivet construction, but you may be able to cut apart the plastic to get to the juicy bits inside. On rare occasions, the toy is constructed in a way that defies disassembly. You can use a hacksaw to rip it apart.

Figures 1-20 through 1-22 show some typical dollar-store toy finds: a motorized vehicle with wired remote, a miniature skateboard (note the additional wheel sets), and replacement wide-tread wheels from a small toy car. The wide-tread wheels are ideal for use as casters on small (6" diameter and smaller) robots. The figure shows the wheels on a 5" diameter tabletop robot base.

Where to Get Stuff: Mail Order

Raw or ready-made, even the hardest to find materials are but a postage stamp away. There is no limit to what you can buy mail order—at one time, they even sold houses through the mail! So, rather than tell you what to look for in mail order, it makes more sense to remind you of the different kinds of mail order.

Figure 1-20 Dollar-store wired RC vehicle

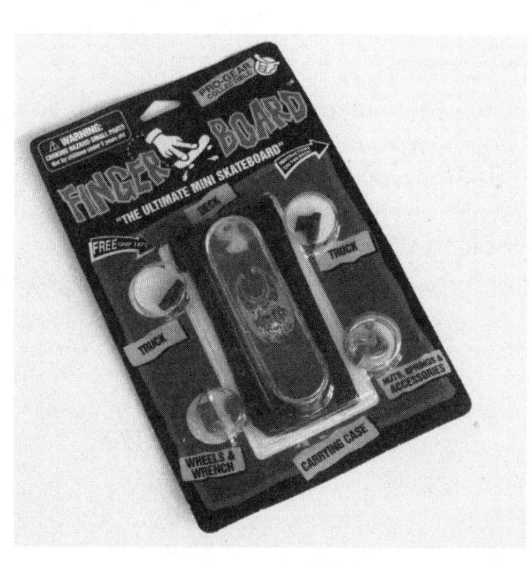

Figure 1-21 Dollar-Store miniature skateboard (note the small wheels)

(And yes, I know that these days, less and less mail order actually goes through the mail—it's a term of convenience.)

- **Retail catalog sales** Some retail stores also offer goods via a separate mail order branch. If the store has a printed catalog, obtain a copy and look through it for special products not carried in the bricks-and-mortar store.
- **Mail order catalogs** This is the typical form of mail order, where a company sends you a catalog or brochure and you order from it. More and more outfits are conducting this kind of business from the Internet, saving the cost of printing catalogs.

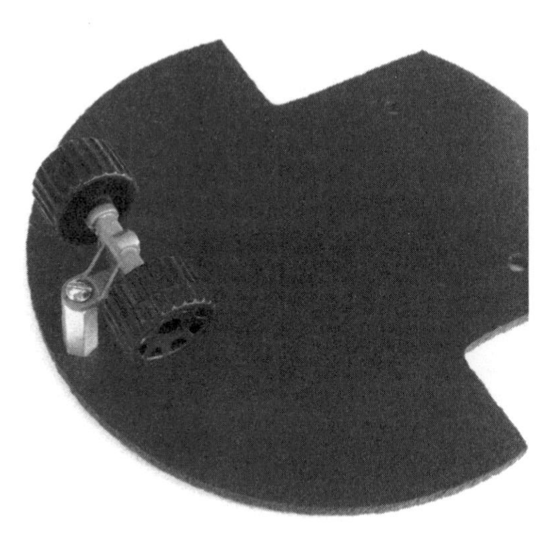

Figure 1-22 Dollar-store replaceable vehicle wheels

- **Internet** Thanks to the Internet you can now find the most elusive part for your robot. The Internet is an extension of catalog sales, where Web pages take the place of printed pages. A disadvantage is that even the best ecommerce shopping cart is not as handy as a nice color catalog that you can read at your leisure in any part of the house.
- **Online auctions** Typified by eBay, online auctions provide a means for individual buyers and sellers to trade. Goods are shipped via the mail.

When shopping via mail order, always compare prices of similar items offered by various companies before buying. Consider all the variables, such as the added cost of insurance, postage and handling, and COD fees.

Where to Get Stuff: Retail

There aren't many robot-building stores at the mall, but you can do quite well at the following retail establishments around town:

- **Hobby and model stores** Ideal sources for small parts, including lightweight plastic, brass rod, servo motors for RC cars and airplanes, gears, and construction hardware.
- **Arts and crafts stores** Sells supplies for home crafts and arts. As a robot builder, you'll be interested in just a few of the aisles at most craft stores, but what's in those aisles will be a godsend!
- **Hardware stores and builder's supply outlets** Fasteners (nuts and screws), heavy-gauge galvanized metal, and hand and motorized tools.

- **Electronic specialty stores** There's RadioShack, of course, which continues to support electronics experimenters. Check the local Yellow Pages under *Electronics-Retail* for a list of electronic parts shops near you.
- **Surplus stores** Electronics surplus stores specialize in new and used mechanical and electronic parts (these are not to be confused with surplus clothing, camping, and government equipment stores). Finding them is not always easy. Start by looking in the Yellow Pages under *Electronics*, and also under *Surplus*.
- **Sewing machine repair shops** Ideal for small gears, cams, levers, and other precision parts. Some shops will sell broken machines to you. Tear the machine to shreds and use the parts for your robot.
- **Auto parts stores** The independent stores tend to stock more goodies than the national chains, but both kinds offer surprises on every aisle. Keep an eye out for things likes hoses, pumps, and automotive gadgets.
- **Auto repair garages** More and more used parts from cars are being recycled or sent back to a manufacturer for proper disposal, so the pickins at the neighborhood mechanics garage aren't as robust as they used to be. But if you ask nicely enough, many will give you various used parts, such as timing belts, or offer them for sale at a low cost.
- **Junk yards** Old cars are good sources for powerful DC motors, used for windshield wipers, electric windows, and automatic adjustable seats (take note though: Such motors tend to be terribly inefficient for a battery-based 'bot). Or how about the hydraulic brake system on a junked 1969 Ford Falcon? Bring tools to salvage the parts you want. And maybe bring the Falcon home with you, too.
- **Lawn mower sales and service shops** Lawn mowers use all sorts of nifty control cables, wheel bearings, and assorted odds and ends. Pick up new or used parts for a current project or for your own stock.
- **Bicycle sales and service shops** Not the department store that sells bikes, but a *real* professional bicycle shop. Items of interest: control cables, chains, brake calipers, wheels, sprockets, brake linings, and more.
- **Industrial parts outlets** Some places sell gears, bearings, shafts, motors, and other industrial hardware on a one-piece-at-a-time basis.

Building a Robot from a Kit

Here's a final topic to close out our first chapter: Robot kits offer an ideal way to learn about the science of robot building. Rather than having to gather all the bits and pieces yourself, a kit lets you concentrate on the building and programming aspects. You don't need to take numerous trips to the hardware store, and—

depending on the kit—you don't even need to pick up a saw, sander, or drill. (Of course, many robot tinkerers enjoy this aspect the most. It's all in your perspective.)

If you're looking to explore the world of amateur robotics with a kit, here are some pointers to keep in mind as you decide which one is best for you. Obviously, cost is a consideration, with most kits falling into one of the following three categories:

- At the low end, costing from $20 to $100, is the kit for a basic nonprogrammable or simple programmable robot. Examples in this category are BEAM robots. These make for good starter kits, especially for younger robo-builders. Specialty online retailers, such as Budget Robotics, specialize in low-cost starter kits.
- The middle ground is the $100 to $300 kit. The Parallax BOE-Bot (the BOE stands for *Board of Education*) usually comes with a more sophisticated means of programming. These kits are perfect for junior high and high school robotics studies.
- The high end comprises specialty kits, such as the walking robots from Lynxmotion or the heavy-duty platforms from Zagros. Prices may vary from a low of $300 to several thousand dollars. They are intended for the serious robotics hobbyist or for education.

Once you've decided on the price range that suits you, the next step is comparing features. You can judge features based on what the kit comes with (two motors, two wheels, etc.) or on what the completed robot does. If you are looking to learn behavior-based programming in robotics, a robot kit that does not permit you to change its built-in programs will not be of much use to you, no matter what hardware it comes with.

Finally, if possible, ask for an electronic copy of the instruction manual that comes with the kit. Many that I've seen are poorly written, and understand difficult to. It can be frustrating enough troubleshooting a belligerent robot; you don't need poor Yoda-like documentation worsening the situation.

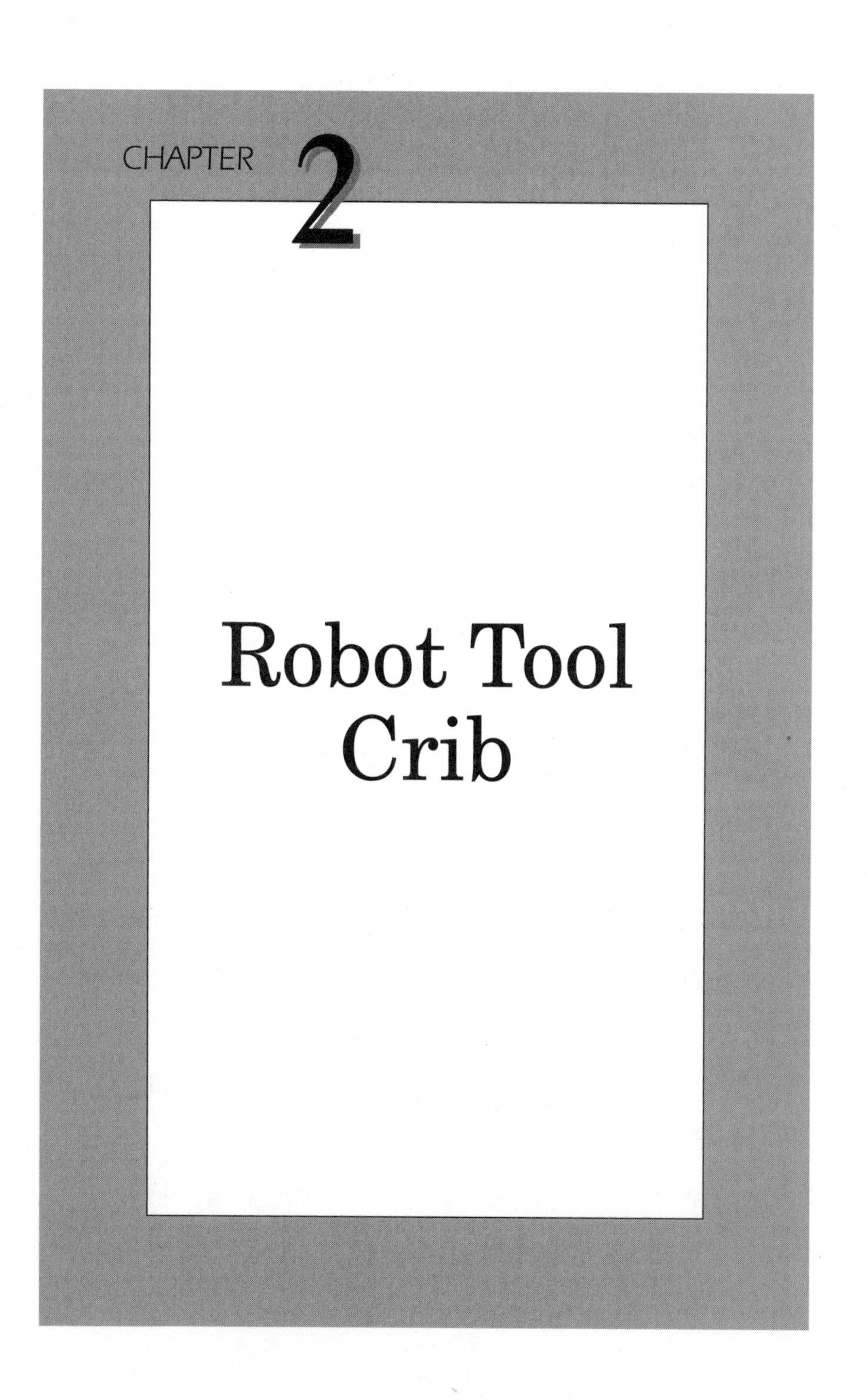

CHAPTER 2

Robot Tool Crib

It's hard to decide which is more fun: building robots or just playing with the tools used to build robots. In any case, I'll let you decide. While you're thinking about it, here's a review of tools—both required and optional—for the robot-building craft. To make this chapter all the more relevant, included are useful buying hints and tips, and how many of the tools are used in building automatons.

Of course, *many* additional tools are not listed here: the mill, the lathe, the reciprocating hole saw, the scroll saw, the band saw, the welding rig, and literally hundreds of others. This chapter does not attempt to discuss *every* tool you're likely to use; rather it covers those common tools that are both the most affordable and the most practical for the average robot builder. You can think of this chapter as "If Bob Vila built robots, these are the tools he'd probably use."

Basic Tools

The following tools are considered must-haves for any robot workshop. Except for the drill, all are hand operated, and most likely are ones you already own.

Figure 2-1

Glasses/Goggles

It is important to provide eye protection against flying debris. The wrap-around style provides the best protection. Use clear lenses for indoors and shaded lenses outdoors in sunlight. Purchase a pair for everyone who will be in the shop area while tools are in use.

I can't stress enough the importance of using adequate eye protection. I'm fortunate enough to have both of my eyes, but I came close to losing sight in one eye from flying debris. As a result, in my workshop, eye protection is now mandatory for myself and all helpers.

Figure 2-2

Tape Measure

A retractable 6' to 12' unit is most convenient. Graduations in both inches and centimeters can be helpful but not critical. A paper or fabric tape (1 yard long, available at yardage or hardware stores, often for free) can substitute in a pinch but may not be as accurate.

Augment the tape measure with one or more wooden rulers purchased at an office supply store or dollar store. Cut them to length as needed and mount them on tools or workbench.

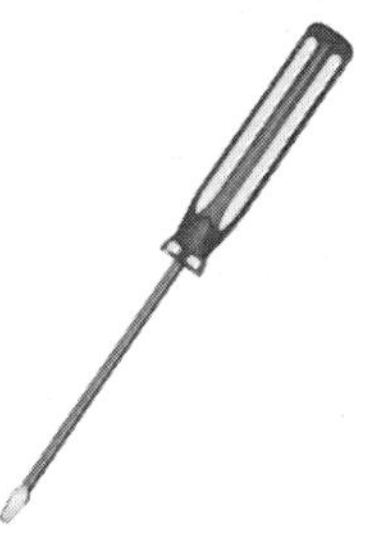

Figure 2-3

Screwdrivers

A set with slot and Phillips in different sizes is preferred. At a minimum, get #1 and #2 Phillips, and medium- and small-tip slotted drivers. Magnetic tips are handy, but not necessary.

Purchase a good set and test the grips for comfort. The plastic of the grip should not dig into your palm. Try soft (rubber) coated grips for extra comfort.

Miniature fasteners (those 4-40 and smaller) require smaller screwdrivers. Purchase a set of #00 and #0 Phillips, and small-bladed, slotted *jeweler's* screwdrivers.

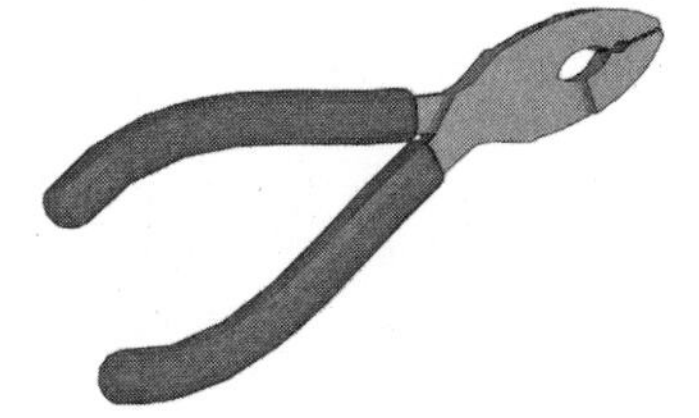

Figure 2-4

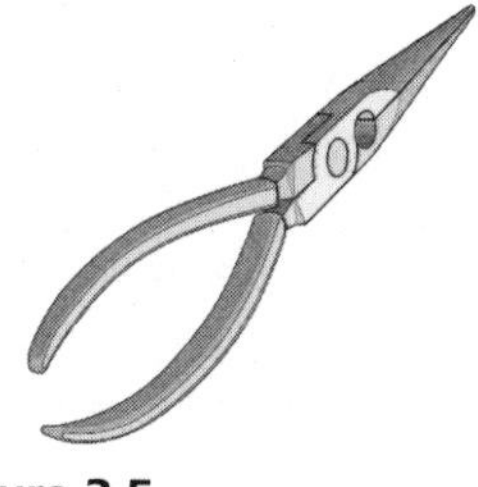

Figure 2-5

Standard Pliers
Needle-Nose Pliers

A pair of each standard and needle-nose pliers is sufficient for 90 percent of all jobs, though avoid using either as a wrench for tightening bolts and nuts (they'll strip the head of the fastener).

For heavy-duty applications, purchase a larger pair of needle-nose pliers. A pair of *lineman's* pliers can be used for the big jobs, and they provide a sharp cutter for clipping nonhardened wire (don't use them to clip steel aircraft cable or music wire).

I once bought a selection of five or six needle-nose pliers, each with a different jaw style (thick, thin, rounded, square, and so on). The special jaw styles are handy every once in a while, but as a whole, I use the standard needle-nose pliers over 95 percent of the time, and the others collect rust. Your mileage may vary.

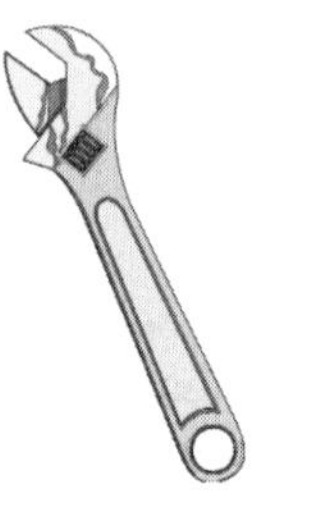

Figure 2-6

Adjustable Wrench

Often misnamed as Crescent wrenches, after Crescent, a popular trade name, adjustable wrenches are used to tighten fasteners of various sizes. One small wrench (opening to about 1/2") and one large wrench (opening to about 1") suffice for most jobs.

Adjustable wrenches are usually sold by handle length, but this isn't as important as the jaw opening size. Cheaper models don't open as wide. A jaw opening of 1/2" is the minimum you should consider; this size fits hex nuts up to 5/16.

The adjustable wrench is one tool where quality counts: Get the *best* you can. Go ahead and spend $10 on a really good wrench. It'll be worth it, and the tool, when used properly, will last your lifetime. (In fact, many of the quality hand tool brands come with a lifetime warranty from the manufacturer. Even if the tool breaks, you can get a free replacement.)

Bargain-basement adjustable wrenches are actually harder to use and can cause damage to fastener heads because they can slip and burr up the metal. Personal tip: Avoid the self-adjusting versions of this tool. I haven't met one that really worked.

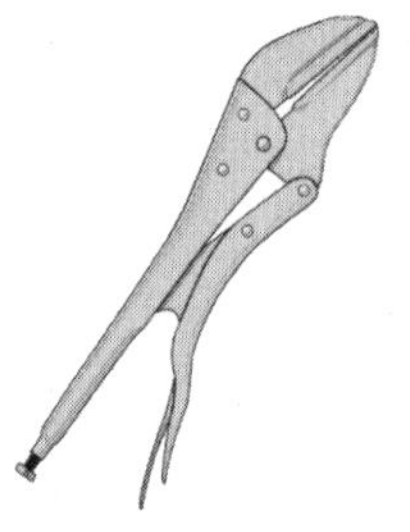

Figure 2-7

Locking Grip Pliers (Vice Grips)

Useful when your hand strength isn't enough. Vice Grips is the most common brand and is recommended for best quality. I bought several pairs of Vice Grips when I was a young man. I'm not so young anymore, but my Vice Grips are still like new. (And no, they didn't pay me to say this! I just like good tools.)

Locking pliers come in a variety of shapes, styles, and sizes. For tight budgets, get one medium-size pair in the style shown. A smaller pair, as well as a pair with needle-nose jaws, can be handy, but are not critical purchases.

A word of caution: It's very easy to overtighten the jaws of the pliers and wreck the material you're working with. Tighten the jaws only so that they firmly hold the work piece and no more. When squeezing the handles of the pliers the jaws should lock around the work piece with minimal effort. If you need to squeeze very hard, the jaws are too tight.

Figure 2-8

Water Pump Pliers

Also known as slip-joint pliers, this wrench is useful for gripping a variety of work pieces. Avoid using this tool for tightening fasteners; the wrench can cause damage to fastener heads. As with adjustable wrenches mentioned previously, purchase the best quality you can. Cheap versions slip, causing damage to the work and injury to you. I can show you the scars on my knuckles if you're interested . . .

PERSONAL TIP: Avoid the self-adjusting version of this tool as well.

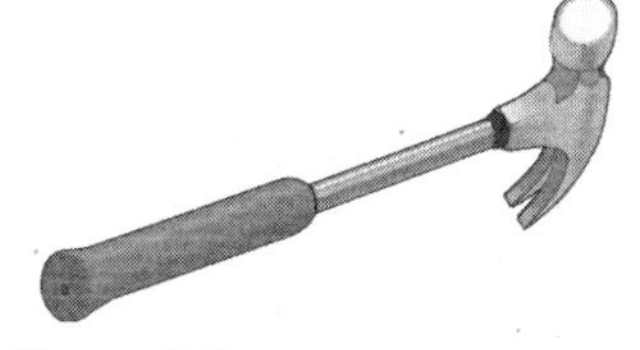

Figure 2-9

Hammer

Surprisingly, this tool is not often used in building robots except to whack parts to shape or to demolish a robot that doesn't behave.

Though most robots are not made with nails, a standard 16-ounce hammer is a useful tool for any workshop.

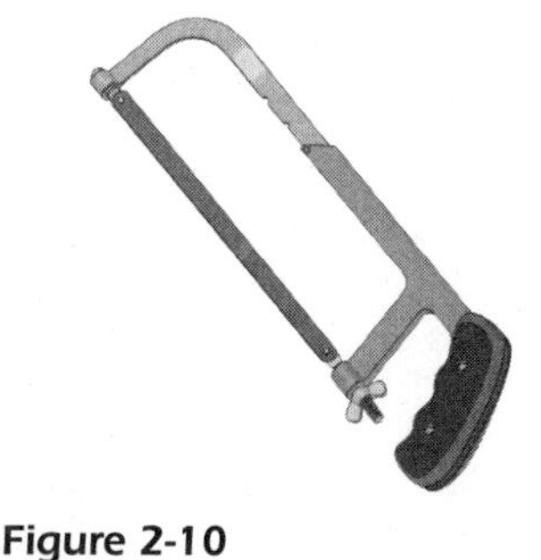

Figure 2-10

Hacksaw

A mainstay of robot building. Purchase at least one hacksaw, if not several, with sturdy metal frames. Look for models that allow quick blade changes yet hold the blade securely.

Common blade sizes are 10" and 12" in length and many hacksaws are adjustable. The smaller size is recommended when working with metal because the short blade gives you more control of the tool. Purchase an assortment of carbide-tipped blades in 18 and 24 *teeth per inch* (tpi). If you have several hacksaw frames, keep blades of different tpi in each one. Even good quality hacksaw frames are fairly inexpensive.

WORK TIP: Select a blade tpi so that a minimum of three teeth, but not more than five or six teeth, engage the work. If fewer teeth engage, cutting will be tough and coarse. If more teeth engage, chips from the work will gum up the blade and dull it. As a general rule, soft or heavy jobs need a coarser tooth pitch; harder, thinner materials need a finer tooth pitch.

Periodically inspect the teeth of the blade, especially if you're cutting metal. If teeth are missing, replace the blade. Trying to cut with a damaged blade only makes you work harder to compensate for the ineffectiveness of the saw.

By convention, hacksaw blades are inserted so that the teeth face forward. Therefore, the saw cuts when you push the blade away from you. However, no strict rule requires you to load the blade in this way. Experiment to see what suits the materials you are cutting. You may find that you have better control over the saw when cutting small pieces if the blade is mounted with the teeth facing you (so the saw cuts when you pull the blade toward you).

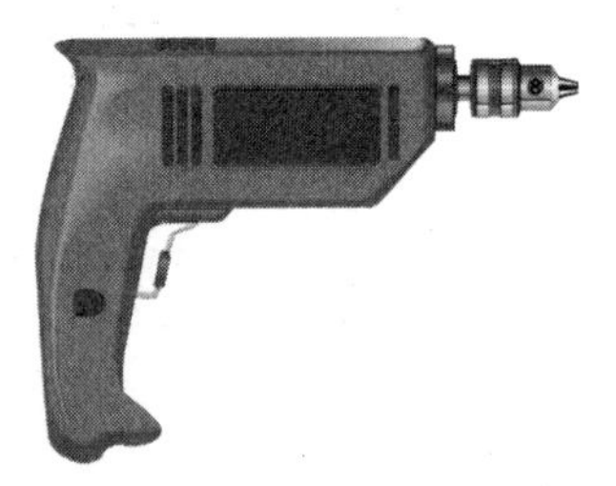

Figure 2-11

Electric Drill

An electric motorized drill with a 1/4" or 3/8" chuck is preferred, but a hand-operated drill will suffice for limited work. The chuck size determines the maximum diameter for the shank of the bit. The vast majority of work on small robots will require bits of 1/4" or smaller.

For greater flexibility, opt for the 3/8" chuck; the big 1/2" chuck is seldom needed, because many larger drill bits use *stepped* shank bits, where the shank is 1/4" or 3/8", even if the bit itself is larger.

WORK TIP: Spring for an adjustable-speed, reversible drill. The slight added price is well worth it.

My father-in-law, a carpenter by trade, taught me an important habit when using a drill. When mounting the bit into the chuck, don't tighten from one point of the chuck only. The chuck has three holes for inserting the key. For a sure fit, tighten from at least two of the three holes. This allows the chuck to firmly grip the bit, and you don't have to bear down on the chuck key as much. Admittedly, this is old-fashioned advice, but over the years I've found it to yield good, consistent results.

This doesn't apply to drills with keyless chucks. These tighten by rotating the chuck (by hand) clockwise or counterclockwise.

Also, be sure to tighten the chuck only around the smooth shank of the bit. Don't insert the bit so far into the chuck that it tightens around the flutes of the bit. Damage to the bit and chuck could otherwise result.

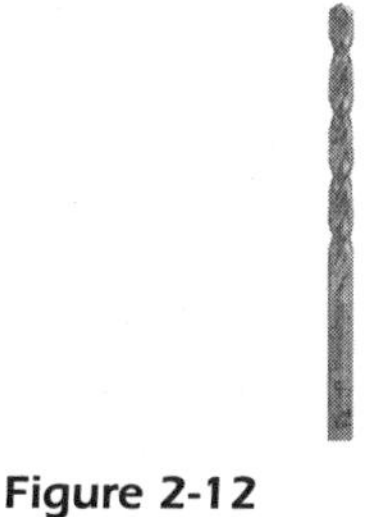

Figure 2-12

Figure 2-13

Drill Bits

Purchase a variety, in so-called jobber length. Tungsten carbide-tipped and cobalt steel bits last longer, but are harder to resharpen at home. A basic set of steel twist bits will last many years when properly cared for. You can sharpen these on a grinder or purchase an inexpensive drill-sharpening tool.

The typical fractional drill bit set contains 29 bits, in sizes from 1/16" to 1/2", in 64ths-of-an-inch steps. For most robotic creations, you'll use only a third of these, but it's nice to have the full set in case you ever need the others.

For the ultimate in long-lasting tools, purchase a set of titanium-coated (technically titanium-nitride-coated) drill bits. These bits are said to last up to six times longer than standard high-speed drills.

An advantage of titanium bits is that they run cooler than standard steel bits. Still, when drilling steel it's a good idea to squirt on a little cutting oil to make the job go easier. Even without the oil, the bits don't expand as readily, so the holes you drill are more accurate. You can get this oil at any hardware or tool store that sells quality drill bits.

Standard twist bits can be used for wood, metal, and many plastics, and have a 118-degree point. Steeper angle bits (a 135-degree split point) are used to drill into hardened and rounded stock. Specialty bits are available for drilling larger holes in wood (also shown here) and for drilling in acrylic plastic without cracking it. These bits use a modified tip and flutes to facilitate drilling into plastic.

BUYING TIP: Purchase a standard 29-piece fractional drill bit set in standard high-speed steel; then augment that set with specific sizes of more expensive longer-lasting bits.

Figure 2-14

Rasps and Files

Rasps are used to remove large amounts of material from the work; files are for finish work. There are an estimated 3,000+ shapes, sizes, and styles of rasps and files, so it's best to purchase a general-purpose set as a starter. A typical set includes one round, half-round, square, triangle, and flat file, in lengths from 5½" to 10".

Note that files are available for either woodworking or metalworking; they're different enough that you'll want to pick one over the other. Select a set based on the material of your robot. Plastics can make use of most metalworking files. Clean your tools regularly to remove any plastic material caught in the file teeth using a metal file brush.

PERSONAL TIP: *Jeweler's files are miniature versions of standard-size workshop files. Sets are available with flat, half-round, round, and triangular files, and are available from specialty tool outlets. I use my jeweler's files far more than the shop files.*

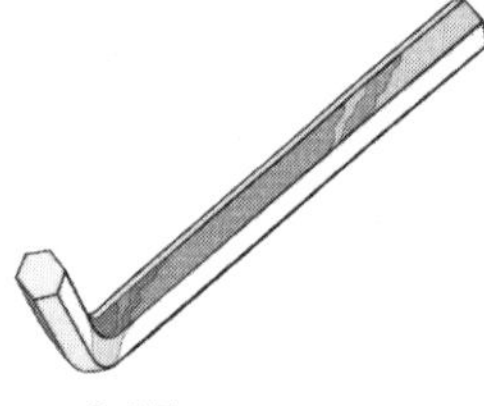

Figure 2-15

Hex Wrenches

Also called Allen wrenches, after the company that helped popularize them, hex wrenches (or hex keys) are used with specialty hex head fasteners, particularly miniature set screws. Wrenches are available in fractional inch and metric sizes. You can purchase hex wrenches separately in various sizes, or in sets.

PERSONAL TIP: I find it difficult to use the all-in-one hex wrench where the individual wrenches cannot be separated from the set. Though you might on occasion misplace an individual wrench, separate tools are far easier to use, especially in cramped spaces.

This said, if you own both metric and fractional inch sets, keep them separate. The wrench sizes between the two are similar but not identical, and it's easy to accidentally use the wrong size of hex wrench, leading to damage to the wrench and/or fastener.

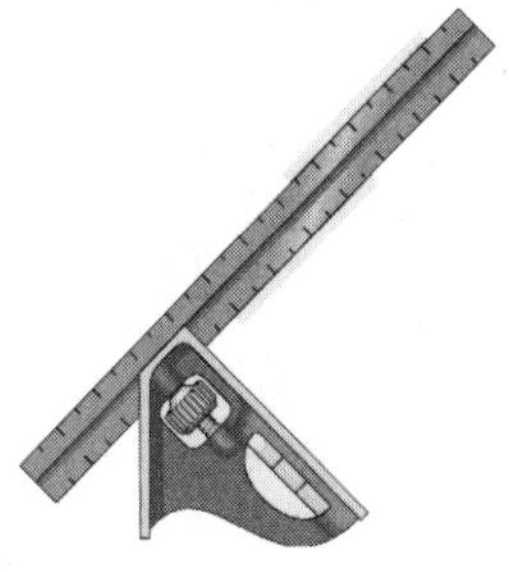

Figure 2-16

Adjustable Square

Squares are used for aligning frames and parts in robots. A basic, adjustable 12" square is satisfactory for the job. Purchase a square with a metal, and not plastic, rule. Some models come with one or more bubble levels; this isn't an important feature.

WORK TIP: Keep the rule lubricated with thin machine oil (such as 3-In-One household oil). The film of the oil will inhibit rust.

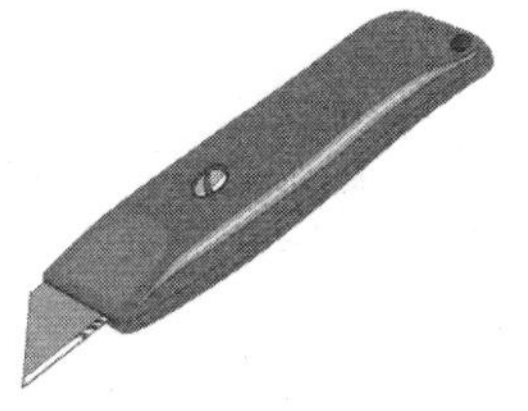

Figure 2-17

Razor Knife

Sometimes called a carpet knife, a razor knife with a replaceable blade is used to cut thin plastic, rubber, paper, and other materials. Get the kind where the blade retracts into the handle for safekeeping.

When cutting sheet material, be sure to back it with a soft surface, or else the blade will become dull, and you'll mar the floor or table. A piece of 1/4" foam board does nicely.

Electronic Construction Tools

These are tools of the trade for electronics construction. Of the tools listed here, only the soldering pencil and meter (either digital or analog) are truly required. The rest are optional.

Not listed are several hand tools commonly found on the electronics workbench. These include the wire stripper and wire cutter, both of which are fairly straightforward and need little discussion here. Buy 'em as you need 'em.

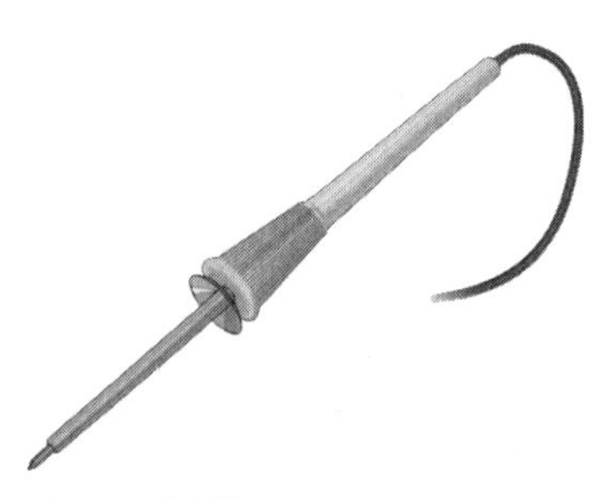

Figure 2-18

Soldering Pencil

This tool is used for all routine soldering of electronic and electrical parts. Do not use a soldering pencil of more than 30 watts or the extra heat may damage components. Most soldering pencils come with a tip that's too broad for fine electronics work. Get a needle-point tip suitable for circuit boards.

Ungrounded (two-wire) soldering pencils are not recommended for digital circuitry. Purchase a grounded (three-wire) soldering pencil, and be sure the electrical outlet is also properly grounded. Get a soldering pencil with a replaceable heat element. You'll save money in the long run.

PERSONAL TIP: Though it adds $30 to $50 to the price, a soldering station with an adjustable heat setting is recommended. Smaller tips need less heat; by using a lower heat, the tip will last longer.

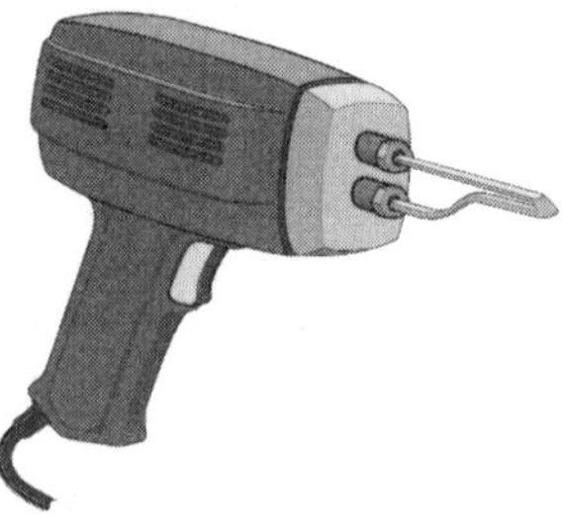

Figure 2-19

Soldering Gun

Use an instant-on soldering gun when increased wattages are necessary for heavier duty jobs. This includes soldering large tabs on power terminals or soldering metal parts together. Most soldering guns are about the same and provide quick on-off heat at the press of a trigger.

Remember: Don't use these for delicate electronics. Soldering guns produce far too much heat and will ruin many types of electronic components, particularly integrated circuits and transistors. Use a low-wattage (25 to 30 watt) soldering pencil instead.

Be sure you can get replacement tips for the gun you purchase—better yet, buy several while you can in case the gun is discontinued and its tips become hard to find.

Figure 2-20

Analog Meter

An analog meter is used to measure *direct current* (dc) and *alternating current* (ac) voltages, amperage, resistance, and continuity.

Although looking old-fashioned, analog meters are sometimes preferred, as the needle is easier to read under different lighting conditions. Many analog meters are also capable of measuring higher amperages than their digital cousins read.

Though only a minor advantage, nearly all analog meters do not require a battery to measure voltage and current (a battery is needed for testing continuity and resistance). This can be beneficial if you are out in the field, and the battery to the meter goes dead. Unlike a digital meter (see the following), you can continue to use your analog meter for important measurements.

Figure 2-21

Digital Meter

Used to measure dc and ac voltages, amperage, resistance, and continuity, many models also test capacitors, diodes, and transistors.

A handy feature found on nearly all digital meters is the audible continuity tester. If resistance is at or near zero ohms, the circuit is assumed to be closed, and a beep indicates continuity. Be sure the meter you get has this helpful feature.

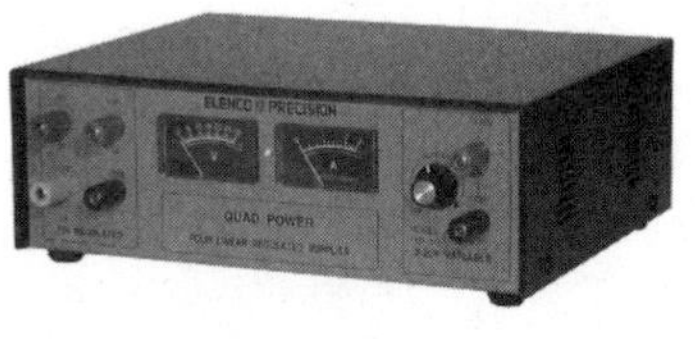

Figure 2-22

Bench Power Supply

Use a bench power supply to provide consistent and clean power to your circuits as you develop them (once you're sure they work, you can run them off batteries).

For power supplies with nonadjustable outputs, a model with +5 and +12 volts dc will suffice. Each output should support at least 1 amp of current, if not several amps. (You will need 3 to 4 amps, at +5 or +12 volts dc, to test most robot motors.)

For power supplies with adjustable outputs, 2 to 12 volts dc is the recommended minimum. Ideally, the power supply should be equipped with its own meter that accurately displays the output voltage. If the power supply lacks a meter, you will need to use a separate volt-ohm meter to monitor the voltage output.

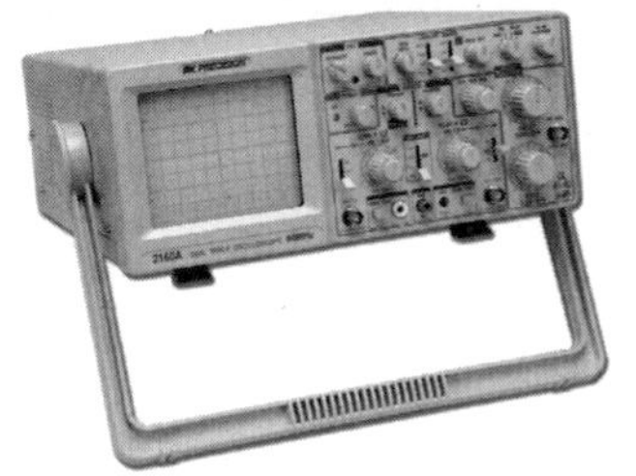

Figure 2-23

Oscilloscope

Definitely an optional item, an oscilloscope allows you to "see" an electrical signal. This tool is a must-have for circuit design and troubleshooting, but it is not a requirement for general electronics tinkering. Opt for a model with no less than 30 Megahertz (MHz) bandwidth. Dual trace, digital storage, and delayed sweep features are nice, but not absolutely necessary. You will, however, find a use for these features if your scope has them.

PERSONAL TIP: PC-based oscilloscopes are often less expensive than full scopes but require a computer in your shop. You must balance the need to have a PC nearby just to use your test equipment, and such a setup is not as portable.

Also available are handheld scopes. These tend to be expensive when they are fully equipped with features such as 50 to 100 MHz bandwidth, delayed sweep, and digital storage.

If your budget is tight, try a used scope from eBay or from a reseller of used test equipment. Opt for one that comes with a money-back guarantee unless you are confident you can fix it should it arrive in nonworking order.

Specialty Hand Tools

What follows are hand tools you don't strictly need for building 'bots, but which you'll find are handy for more sophisticated designs.

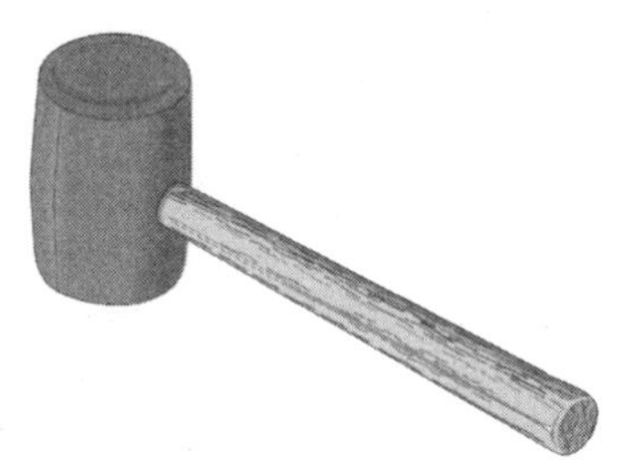

Figure 2-24

Mallet

A standard rubber mallet—nothing fancy. I prefer the two-headed mallet with both hard and soft rubber tips.

Do note that the rubber can become damaged if you try to use the mallet as a hammer. The purpose of the mallet is to *gently* tap parts in order to seat or remove them. Using it as a sledgehammer will definitely shorten the life of the tool.

WORK TIP: Use a small block of wood between the mallet and part you want to loosen or set. The shock imparted from the mallet will transfer through the block and into the part. There is less risk of damage with this method, and the head of the mallet won't get as chewed up.

Figure 2-25

Hot-Melt Glue Gun

This comes in either mini or standard size. For standard size, get one that can use high-temperature (as opposed to low-temperature) glue sticks. A dual-temperature model is the all-around best bet, though these cost more.

The better glue guns accept different tips. The standard round-hole tip is suitable for general glue application. With this tip you can apply a bead of glue that can be spread by exerting pressure on the parts being joined. Additional tips include the spreader applicator, available in different widths. This tip applies a wide coat of glue of a specific thickness and is ideal when laminating materials.

You can purchase replacement glue sticks at hardware and home improvement stores, as well as most arts and craft stores. The latter also offers colored and glitter glue sticks, which you probably won't use much in your robot. Be sure to select the proper temperature glue sticks for your gun. High-temperature sticks will not melt in a low-temp gun, and low-temperature sticks are unnecessarily overheated in a high-temp gun.

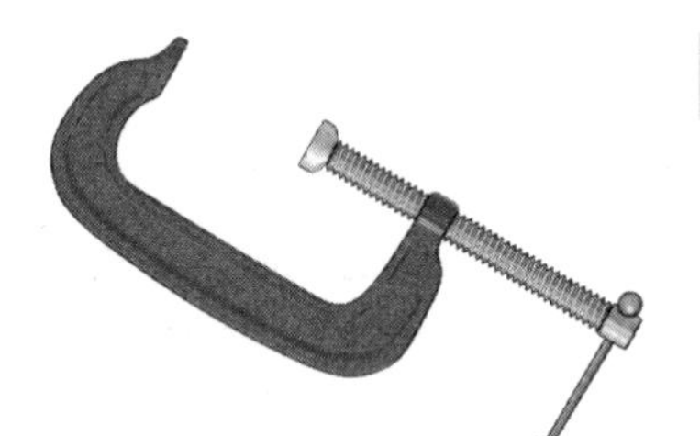

Figure 2-26

Clamps

An assortment of C-clamps for holding pieces is necessary for drilling and gluing. Heavy-duty spring clamps (they look like overgrown clothes pins) are handy for smaller jobs.

Get several different sizes. Use the right size for the job. An oversized spring clamp can exert too much pressure and cause damage to the material you're working with.

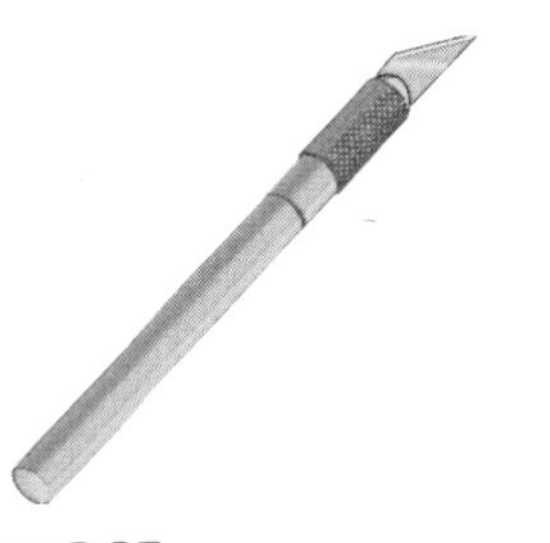

Figure 2-27

Hobby Knife

Interchangeable blades make the hobby knife, such as the X-Acto brand, a handy tool for cutting cardboard, foam board, and thin plastics.

A word of caution: The blades in these knives are *extremely* sharp. Use with care, and never leave the knife out in the open where children may play with it. If the knife does not have a blade cover (or does not retract), purchase a small plastic box for the blade and knife, and keep them safely stored there when not in use.

Though the blades are typically made of stainless steel, they can rust over time and with exposure to the air. If you buy an assortment of blades for now and the future, apply a couple of drops of oil to them, and then wrap the blades in cotton gauze or in a plastic sandwich baggie. They'll last for years this way.

Figure 2-28

Socket and Box Wrenches

Socket wrenches make fast work out of tightening and loosening nuts and bolts. Standard-size sets for automotive work are usually too big for small mobile robots; get a miniature set in sizes more appropriate for the standard robot—2 to 15 mm for metric size, or 2½" for standard sizes.

PERSONAL TIP: Don't skimp on quality. A good set of socket wrenches should last you a lifetime. "Buy once, keep forever" is my motto.

Box wrenches are used for the same job, except for certain brands that don't have a ratchet mechanism. They're useful for tightening nuts and bolts when space doesn't allow for a socket wrench.

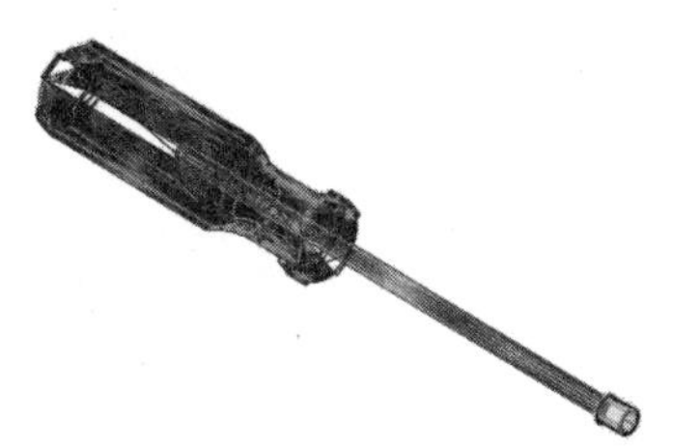

Figure 2-29

Nut Driver

This tool tightens hex head nuts and screws. These look like screwdrivers but are designed for use with hex nuts. They come in metric and standard sizes. On the standard front, common driver sizes are

Driver	For hex nut
1/4"	#4
5/16"	#6
11/32"	#8
3/8"	#10
7/16"	1/4
1/2"	5/16

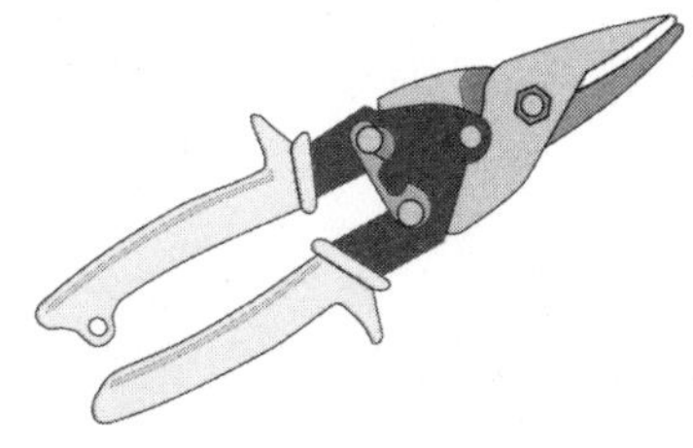

Figure 2-30

Metal Snips

Also called tin snips or aviation cutters, these tools are available in different versions for straight, right-hand, and left-hand cuts. These see use in robotics for cutting sheet thin (to about 24 gauge) sheet metal and up to 1/16"-thick aluminum.

Heavier metal needs a suitable metal sheer, a foot- or hand-operated machine that can cut much thicker material.

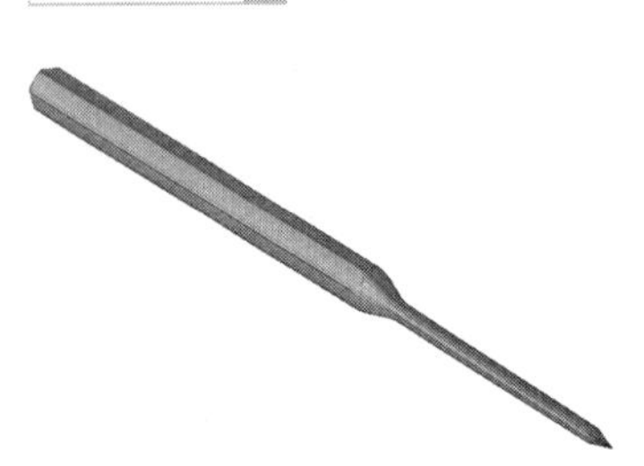

Figure 2-31

Drill Punch

This is used to set a starting mark for drilling in wood or metal (not recommended for acrylic plastic, which can crack). A spring-loaded punch can be used without a hammer.

Use a sharp nail as an alternative. As needed, grind the tip to a sharper point. This will ensure a good mark in the material.

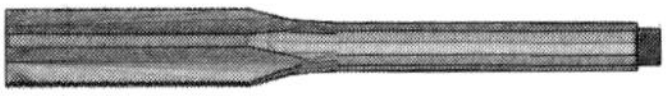

Figure 2-32

Die Thread
Tap and Die Set

Figure 2-33

Taps thread holes for bolts and machine screws; dies thread shafts for nuts. Select a set in standard or metric size, as needed. Don't go cheap; the inexpensive tap and die sets break and dull easily.

Assuming standard sizes, purchase a set with 2-56 to 1/4" *unified national coarse* (UNC) threads—or purchase individual taps and dies as needed. These can also be used for thick plastics and non-ferrous (aluminum or brass) metals.

When tapping metal, even aluminum, it's always advisable to use tap threading oil for lubrication (the same oil can be used with dies). The oil helps reduce overheating and dulling of the tool and makes the work much easier for you.

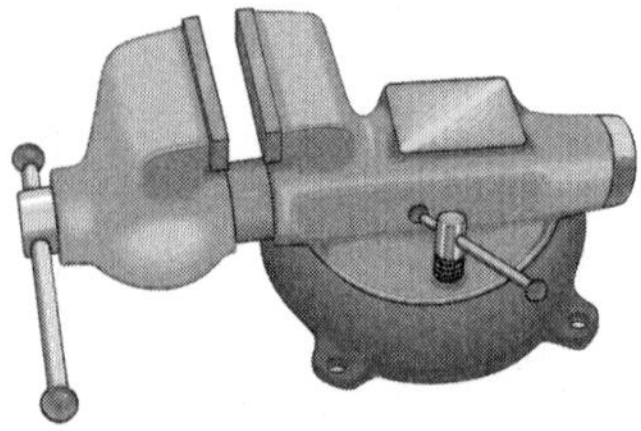

Figure 2-34

Table Vice

A vice is used to hold pieces secure while you cut and drill. A jaw opening of up to 4" is enough for most jobs.

PERSONAL TIP: Periodically apply a lightweight machine oil (3-In-One) to the surfaces of the vice; then wipe it off with a clean paper towel. The light film of oil inhibits rust.

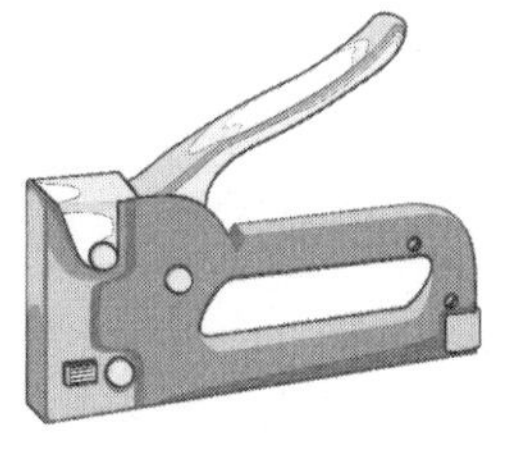

Figure 2-35

Heavy-Duty Stapler

Use this to attach rubber or thin plastic to wood-framed robots. Boxes of staples, in different sizes, are available at any hardware store.

Now, I know what you're thinking: Why purchase a stapler like this when there's a perfectly good paper stapler in your desk? The problem with this line of thought is that the paper stapler doesn't impart the force necessary to drive a staple into thick material, especially PVC plastics. What's more, the staples themselves are thin wires; you want a heavier wire that can penetrate 1/4" or more of material.

Specialty Power Tools

As with the previously mentioned specialty hand tools, these are power tools that are not strictly required, but can greatly decrease the time and energy spent on constructing robots. "Required" is, of course, a subjective thing. Personally, I wouldn't do without any of the tools that follow. I use them all quite regularly.

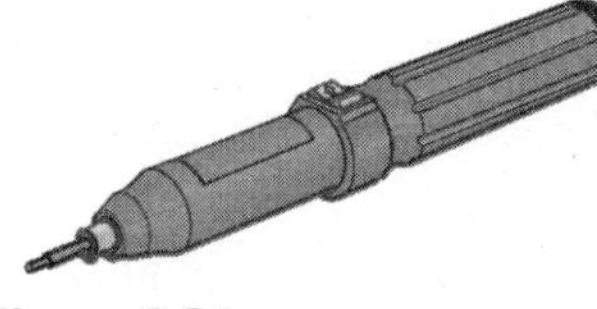

Figure 2-36

Motorized Screwdriver

Rechargeable and reversible models make short work out of most any job. You may want to purchase a couple of units so that you can use one while the other is recharging.

Eventually, your motorized screwdriver will give out, typically because the battery is no longer able to hold a charge. Don't throw the screwdriver away! Instead, hack it for its motor and gear box. Many motorized screwdrivers can be used as is (without the battery compartment) as a drive motor for a robot. Mount the screwdriver to the base of the robot using clamps or even nylon tie-wraps. Wheel shafts or other power transmission components can be made using old power screwdriver bits (see next section) and accessories.

Figure 2-37

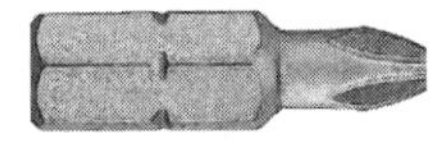

Figure 2-38

Figure 2-39

Phillips
Power Screwdriver Bit
Driver Bits

Most motorized screwdrivers (see previous section) come with an assortment of driver bits. Purchase extra as needed. Available in common drive types: slotted, Phillips, square, Torx, and Pozidriv, each in several sizes.

Also available are drill bits for making quick holes. These are commonly available in small sets, with drill bits from 1/8" to 1/4".

Figure 2-40

Dremel

The ubiquitous motorized hobby tool, the Dremel, can be used as a miniature drill, sander, cutter, or grinder. Kits come with a variety of bits (most you'll never use), and additional bits are available for separate purchase. The 1/8" drill and miniature cutoff saw are among the most practical.

Of course, other brands of motorized hobby tools are available, but Dremel is by far the most popular.

Figure 2-41

Roto Zip

Originally intended for drywall installers, the Roto Zip is technically referred to as a spiral saw. The 1/8" bit is a combination saw and drill. Versions of the bit are available for soft materials, like Sheetrock, as well as plastics and soft, non-ferrous metals.

Note that a similar spiral-cutting drill or blade is available for the Dremel tool and is used in much of the same manner. The Roto Zip is preferable when cutting thicker and heavier materials.

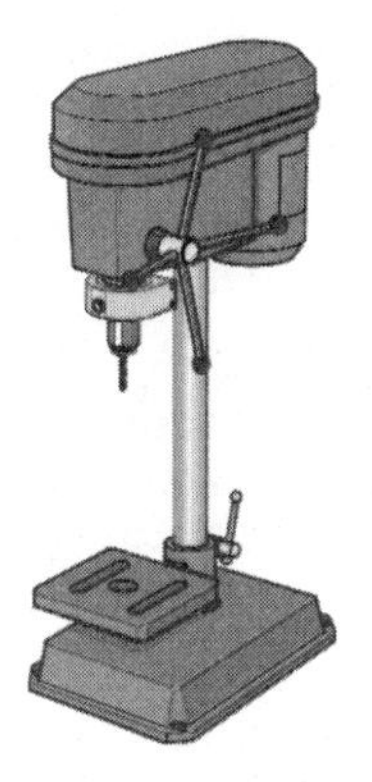

Figure 2-42

Drill Press

A bench or self-standing drill press allows for more accurate holes. There is no need to spend lots of money here; an ordinary $75 to $150 model will suffice. If yours doesn't come with a drill press vice, purchase one separately, probably for about $30, or more if it has a magnetic base.

Nearly all drill presses have variable speeds by adjusting the drive belt. High speeds can be used when working with wood, but slower speeds should be used with plastic and metal. You definitely want an adjustable-speed model.

Handy, but not required, is electronic speed control. Rather than adjust the belt on a set of pulleys, you merely dial in the speed you want to use. This feature is helpful when alternating between different materials, and between various sizes of bits (general rule: the larger the bit, the slower the speed). However, electronic speed control adds to the cost of the drill press.

Figure 2-43

Circular Saw

A standard 7¼" circular saw can be used to make short work out of cutting wood and ¼" or thicker plastic sheets.

Be sure to use the correct blade for the type of material you are cutting. For straight cuts, purchase a saw fence, available at most hardware and home improvement stores.

The strongest circular saw is the worm drive and is intended for cutting heavy lumber, such as the framing of a house. Worm-drive saws are much heavier than the standard models and, unless you're building house-size robots, they aren't really needed. Save your money for other tools.

Figure 2-44

Table Saw

A $7^1/_4$" to 10" table saw provides more accurate cuts than a handheld circular saw. As with the circular saw, be sure to use the correct blade for the material you are cutting.

A word of mild caution is due: Many hardware and home improvement stores sell economy table saws for about $80 to $100. These saws will certainly cut through material, but they are not known for their accuracy, especially when carelessly assembled. Don't trust their guide fence; measure everything yourself with a carpenter's square. Periodically recheck the alignment of the table and make adjustments as necessary to ensure true, square cuts.

Alternatives to table saws include the band saw and the radial arm saw. Each type of saw has its own special place in the workshop. Personally, I prefer a radial arm saw for general cutting, but some jobs are best handled by a good table saw.

Figure 2-45

Orbital Sander

Orbital sanders are designed for finish sanding of wood, but they are also good general-purpose sanders—when equipped with the proper sheet of abrasive—for plastics and even non-ferrous metals. Mount the sander upside down in a vice to use it as a bench sander.

Sand papers are commonly available with grits ranging from very rough (50) to very fine (over 300). Use a lower grit to remove large amounts of material and a higher grit to achieve a smooth finish. You can sand many soft plastics, such as expanded rigid PVC, with a 150- to 175-grit paper. Metal can be sanded using 250- to 300-grit wet/dry paper, used wet. The paper is wetted with ordinary water.

Figure 2-46

Figure 2-47

Jigsaw
Jigsaw Blade

Jigsaws are used to cut intricate shapes in wood, plastic, and thin sheet metal. The jigsaw is basically a motorized hacksaw, but is easier to use for inside cuts (such as cutting a circle out of a sheet of wood). It is critical that you match the blade to the material you are cutting. See the note for hacksaws and blades, earlier in the chapter.

Figure 2-48

Miter Saw

The motorized miter saw is useful for quick and accurate cutting of rod, bar, channel, and angle stock. The saw can make straight cuts or be adjusted to make angle cuts. Miter cuts at 45 degrees are used to build frames.

Use the appropriate blade if cutting ferrous or non-ferrous metals. I prefer a toothed blade rather than an abrasive cutoff blade. When using toothed blades for cutting metal, be sure to always use the appropriate blade lubricant, or else the life of the blade will be significantly reduced.

I make heavy use of my miter saw for cutting channel and angle aluminum for robot frames. The construction time is greatly reduced compared to using a hacksaw and the finished product is much better. The cuts are straight (there are no "rounds" so common with manual hacksaw cuts) and the miter angles match perfectly. There is generally less flash (rough metal edges) to remove with the file and grinder.

I would have placed the miter saw in the must-have category except that not everyone builds robots using aluminum stock. If you do, then by all means consider this an essential tool.

SAFETY TIP: Miter saws can be dangerous when cutting small pieces. Use a clamp to hold the piece on both sides of the blade when being cut. Exercise the usual cautions and use protective eyewear and gloves at all times.

Figure 2-49

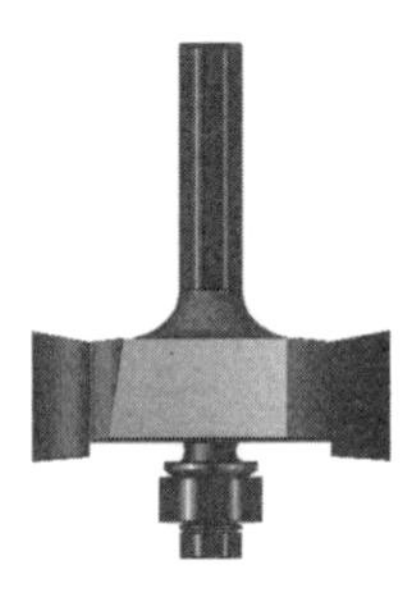

Figure 2-50

Router
Router Bits

The router is used to form special edges and grooves in materials. In the woodworking shop, a prime use of the router is to create joints for cabinets. For robotics, the router can be used to make grooves and raceways for wiring and parts, and to finish the edges of wooden and plastic frame robots.

Most routers come with a small assortment of router bits. Augment this assortment with additional bits as needed.

A note about the successful use of routers: More than any other tool described in this chapter, the router requires the most skill to use properly. If you've never used a router before, don't expect professional-looking results the first time out of the box. Practice with scrap pieces. Routers turn at a very high speed (up to 30,000 rpm), and when used on 1/4"- or 3/8"- thick plastic and wood, the bits do their business very quickly. It takes skill to create the fine cuts and grooves that are the hallmark of the expert.

Figure 2-51

Grinder

The handheld grinder is useful when working with ferrous and non-ferrous metals. It takes the place of hand files for removing burrs and flash.

If you plan on doing lots of metal work, consider investing in a bench grinder. Models that do a decent job are available for under $50. These attach to your workbench and pro vide two replaceable grinding wheels on either side of the tool.

If you work with acrylic plastic, you can outfit one of the grinding wheels with a cloth polishing wheel. Purchase rouge or another polishing compound from a plastics special shop and apply it to the spinning wheel. You may then polish the acrylic plastic to a bright shine.

Circular and Table Saw Types

Blades for circular saws and table saws are available in both all-steel and carbide-tipped versions. All-steel blades are economical, but carbide-tipped will last longer and can be resharpened more times. Also available are diamond blades, but these are considered specialty tools (typically used for tile and masonry), and they can be frightfully expensive.

Though most saw blades are designed for wood (and some for ferrous and nonferrous metals), many can also be used with 1/8" and thicker plastic sheets. Typically, however, the better blades for plastics are those with so-called thin-kerf and thin-rim designs, which limit the heating—and therefore melting—of the plastic.

As a general rule, the harder the plastic, the more teeth there should be on the blade. For example, acrylic and polycarbonate are best cut with a 60- to 80-tooth blade (these numbers are for a 10"-diameter blade; the number of teeth decrease when using a smaller diameter blade). When cutting plastic, it's best to use the largest diameter blade possible to minimize blade heating.

Note that the blade profile illustrations that follow are representative only and can vary by manufacturer.

Figure 2-52

General Purpose

Use this blade when others are not available, or the type of work is too varied to make blade changes practical.

Figure 2-53

Fast-cut Combination

This is recommended for smooth cuts in softwoods and hardwoods (such as oak) in any direction, including miter cuts. One can also be used as a general-purpose blade for many plastics.

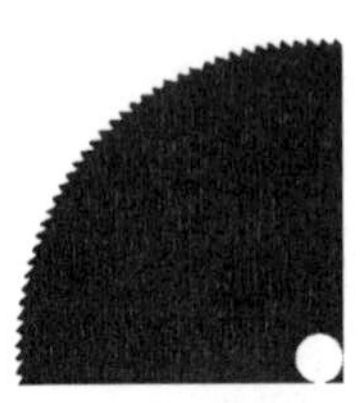

Figure 2-54

Hollow-ground Plywood

This has a special tooth design for construction (softwood) plywoods. When used properly, the finish side of the plywood does not splinter. One can also be used for some plastics. In general, this is a good all-around blade for cutting light sheet materials of wood and plastic.

Figure 2-55

Fine Crosscut or Rip

This blade is most used for crosscuts or rips in all types of wood and is a common choice for general-purpose sawing.

Figure 2-56

Finishing

This provides smooth cuts in wood and other soft materials of all thicknesses.

Figure 2-57

Paneling

This paneling blade allows smooth cuts in thin materials, such as paneling, without splintering. This saw is recommended for Masonite and similar resin-based fiber woods.

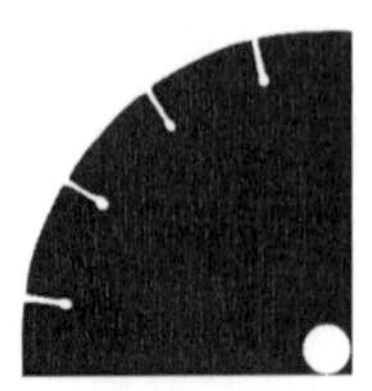

Figure 2-58

Ferrous Metals (Iron and Steel)

This cuts iron-based metal. Lubricating the blade during cutting is usually required.

Figure 2-59

Non-Ferrous Metals (Aluminum, Brass, Zinc, etc.)

This cuts non-iron-based metal. Lubricating the blade during cutting is usually required.

Figure 2-60

Aluminum

This is a specialty blade for aluminum. Lubricating the blade during cutting is usually required.

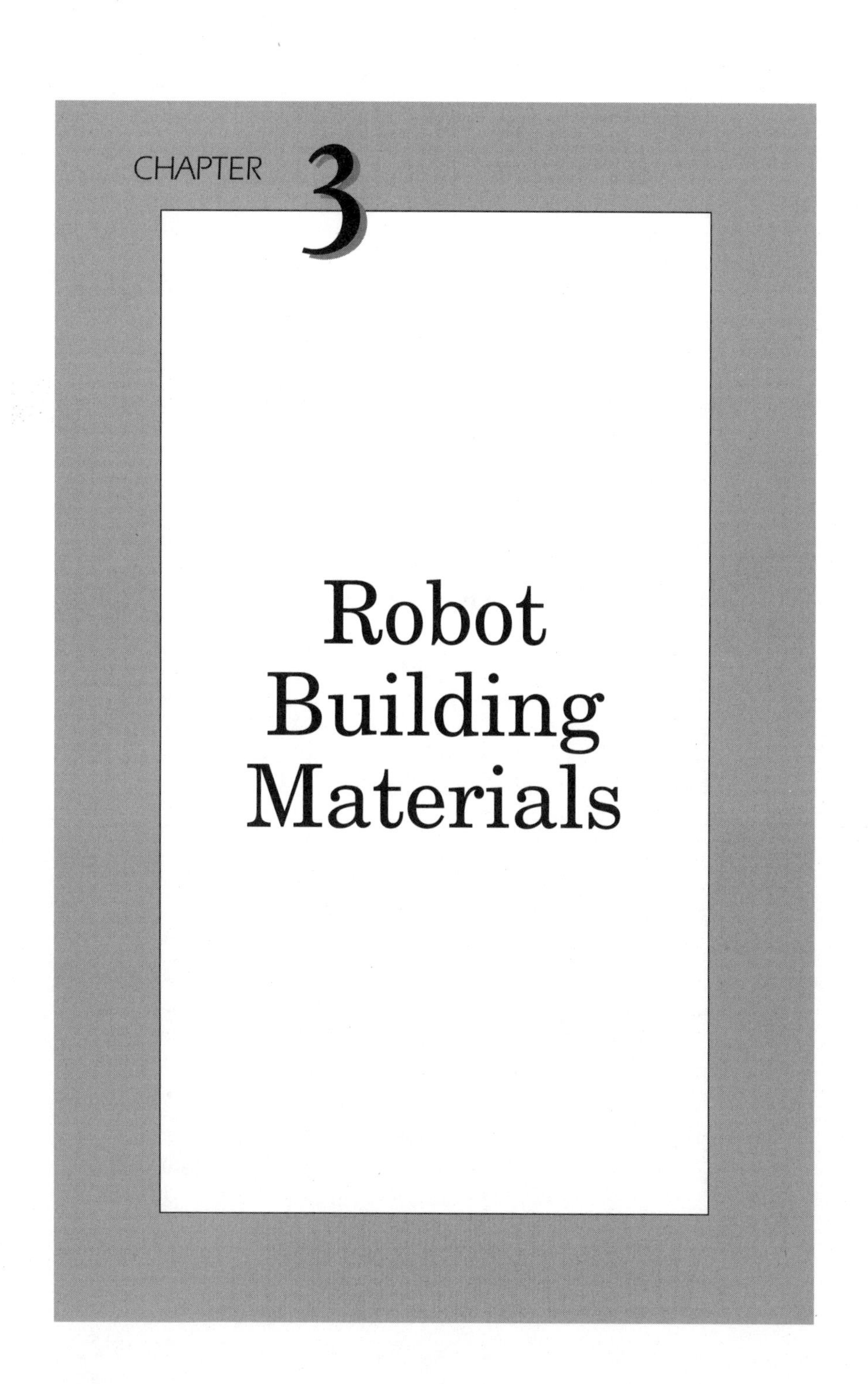

CHAPTER 3

Robot Building Materials

Your choice of materials for your robot's base will help determine not only its strength, weight, and size, but also the level of difficulty of construction, specific tools that may be required, and the skill level needed to achieve acceptable results. Although many robots can be constructed using ordinary wood or plastic, choices abound with even these materials. It also pays off to be aware of the variety of materials that are available to you so that you can select an alternative if the situation calls for it.

This chapter discusses the materials you are most likely to use in your robot-building endeavors. Because the field is a large one, this chapter does not include construction techniques that require specific materials. Chapter 5, "Mechanical Construction Techniques," discusses overall construction techniques; Chapter 7, "Bases and Framing," is devoted to the finer details regarding plastic and metal bases.

Material for Robot Bodies: An Overview

Recall from Chapter 1, "The Basics of Robot Bodies," (you read that chapter, right?) that robots can be constructed from aluminum, steel, wood, plastic, paper, foam, or a combination of them all. Taking this topic further, here are the most common materials used in constructing robot bases and other components. More information about each of these can be found throughout this chapter.

- **Wood** Wood is easy to work with, can be sanded and sawed to any shape, doesn't conduct electricity (avoids short circuits) unless wet, and is available everywhere. However, ordinary construction woods are not recommended; use the more dense (and expensive) multi-ply hardwoods for model airplane and sailboat construction.
- **Plastic** Pound for pound, plastic has more strength than many metals, yet it is easier to work with. You can cut it, shape it, drill it, and even glue it. Effective use of certain plastics requires some special tools. Some unique plastics might be harder to get unless you live near a well-stocked plastic specialty store; mail order is a great alternative.
- **Foam board** Art supply stores stock what's known as foam board, also called foam core. Foam board is a sandwich of paper or plastic glued to both sides of a layer of densely compressed foam. The material comes in sizes from 1/8" to over 1/2", with 1/4" being fairly common. Paper-laminated foam board can be cut with a sharp knife; plastic-laminated foam board should be cut with a saw. Foam board is especially well suited for small robots when low weight is important and for making quick prototype models.

- **Aluminum** Aluminum is a great robot building material for medium and large machines because it is exceptionally strong for its weight. Aluminum is easy to cut and bend using ordinary shop tools. It is commonly available in long lengths of various shapes, but it is somewhat expensive.
- **Steel** Because of its strength, steel is used in the structural frame of larger robots, particularly combat robots. This prime material of industry is, however, difficult to cut and shape without special tools. Stainless steel is favored for precision components, such as arms and hands, and also for parts that require more strength than a lightweight metal (such as aluminum) can provide.
- **Other metals: tin, iron, and brass** Tin and iron are common hardware metals, often used to make angle brackets, sheet metal (various thickness from 1/32" on up), and (when galvanized) nail plates for house framing. (Note that iron is used to make steel, but iron is a softer metal because of its different composition and is often classified separately from steels.) Brass is commonly found in decorative trim for home construction projects and as raw construction material for hobby models. All three metals are stronger and heavier than aluminum. Cost is fairly cheap, hence the popularity of these metals.

Wood

If billionaire Howard Hughes could build the world's largest powered airplane out of spruce, how hard could it be to construct a small robot out of wood? Indeed, wood can be an ideal construction material, especially for robot bases, which need to be sturdy and strong, yet easy to cut and drill.

Many varieties of wood exist, but 90 percent are either too expensive or too exotic (and therefore hard to get) for robotics use. Wood can be broadly categorized as hardwood or softwood. The difference is not the hardness of the wood, but the kind of tree the wood is from. *Hardwood* is produced from trees that bear and lose leaves (deciduous) and *softwood* trees bear needles (coniferous) or do not undergo seasonal change (nondeciduous). In general, deciduous trees produce harder and denser woods, but this is not always the case. A common hardwood that's very light and soft is balsa.

The processing of hardwoods and softwoods into milled products greatly influences their practicality in robotics. Solid woods, like that used for lumber or nonengineered flooring, aren't as strong as plywoods; the latter is made by laminating layers of thinner sheets of woods together (see Figure 3-1). Wood has a grain, and its strength is in the direction of the grain. With plywood, the grain structure is turned at right angles for each layer. This greatly strengthens the

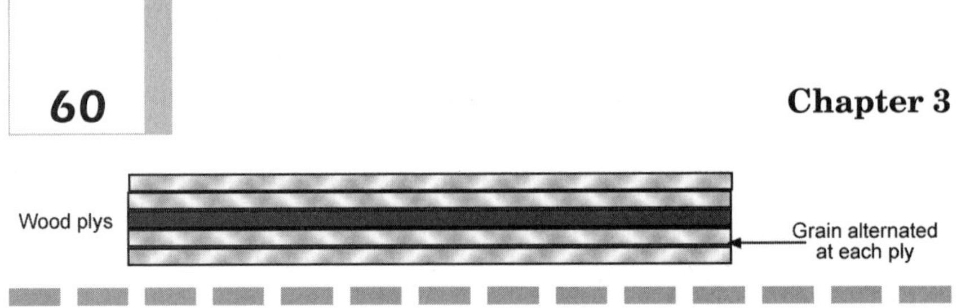

Figure 3-1 Cross-section of plywood shows how the material is made by laminating several thin sheets together.

wood in all directions. It also allows the wood to be produced in thinner sheets yet have good strength properties.

Construction plywoods, such as the ones used for building homes, can contain hardwood or softwood, or a combination of both. The inner plys may be a softwood, and the outer plys a hardwood—or some variation thereof. These have marginal use in robotic bases. A better option is the all-hardwood plywood, a mainstay of the model shipbuilder. Also available is aircraft-grade plywood. Both are very dense and strong, yet they can be readily sawn and drilled. Birch and spruce are commonly used for these plywoods. Because of their cross-laminated construction, plywoods also resist warping.

The density of wood determines its weight per cubic inch or centimeter. A heavy wood, such as oak, is not desirable for mobile robots because of the additional weight, despite oak being a very sturdy wood. Conversely, although pine is light, it is a softwood and is not very strong. You'd need a fairly thick piece (at least 1/2") to provide adequate dimensional stability for a robot base, arm, or other mechanical component.

Best Types of Wood for Robotics

Although there are thousands of types of trees (and therefore wood) in the world, only a relatively small number of them share the following traits that make them ideal for building robots:

- Relatively easy to find—if not locally, then from a variety of mail-order outlets
- Reasonably priced (up to $10 is reasonable, considering the small amounts needed for most robot projects)
- Strong for its weight and size
- Readily drilled and sawn using ordinary shop tools

Let's delve deeper into the best woods for robots.

Wood Classifications As stated above, wood can generally be classified into two groups: hardwood and softwood. Softwoods tend to be low in density, meaning that they require a certain bulk in order to achieve a desired strength.

Plank or Ply Unless you're Abraham Lincoln, building a robot from lumber hewn from the forests of Kentucky, you'll most likely purchase milled wood in either plank or laminate (ply) form:

- *Planks* are lengths of wood milled from the raw lumber stock. They are available in standard widths and thicknesses and come in either precut lengths (usually 4, 6, and 8 feet), or are sold by the linear foot.
- *Plywood*, as noted previously, is made by sandwiching one or more types of wood together. The grain is alternated at each ply for added strength.

Both plank and ply are made from softwoods and hardwoods. Depending on where you live, your local home improvement store is likely to have mostly softwood ply and maybe just a few types of hardwood plank (mostly oak). For a wider variety you need to shop at a hardwood specialty store or by mail order.

Hardwoods for Robots Refer to Table 3-1 for a list of suitable woods for robotics. No differentiation is made between blank and ply because they are often available in either form. I usually prefer using plywood, because the material is available in a wider variety of sizes and thicknesses. In the descriptions for the various woods, the term "work" means to drill, saw, or turn using hand or motor-driven tools.

Do note that the availability of wood is regional. Not all of the woods are readily available everywhere. Some are even protected woods and are available only in limited quantities. All of the woods in Table 3-1 are hardwoods.

More High-Strength Woods The following woods are known for their strength, but because they are also dense, they are quite heavy for their dimensions. If weight is not a critical issue for you, one of these high weight-to-strength woods may be appropriate for your application.

Of the following woods, all except yew are hardwoods. Note that as a general rule, the denser the wood, the harder it is to drill and cut. There are exceptions, of course: Pear is a high weight-to-strength wood that is easy to turn in a lathe because of the construction of its grain.

- **North America** American beech, boxwood, dogwood, hickory, honeylocust, mesquite, oak, pear, persimmon, sugar maple, yew
- **Central and South America** Brazilwood, cocobola, degame, kingwood, pau marfim, purpleheart, rosa peroba, rosewood, tulipwood
- **Europe** Beech, boxwood, European hornbeam, oak, pear, yew

About Low-Strength Woods Woods that exhibit low strength are generally not advisable for use in robotics except for special application as noted in the following list. Low-strength wood, such as balsa or basswood, may also be lightweight but may break easily under stress.

Table 3-1 Hardwoods

Common Name	Latin Name	Source	Description	Qualities
Ash, white	Fraxinus americana	North America	White ash is one of the best hardwoods for robots. It's readily available in the U.S. and Canada, is low cost, and has an excellent weight-to-strength ratio.	Weight: Medium Hardness: High Strength: High Working: Medium
Guarea	Guarea cedrata (also G. thompsonii)	Africa	This attractive wood has a low weight for its relatively high strength.	Weight: Low Hardness: Medium Strength: High Working: Medium
Muninga	Pterocarpus angolensis	Africa	Muninga is a coarse-grain wood that is easy to work with and is somewhat soft, yet is very strong.	Weight: Medium Hardness: Low Strength: High Working: High
Birch, yellow	Betula alleghaniensis	Canada, Eastern U.S.	Yellow birch is strong, and stiff. Because its hardness is rather low, it is often used for "turning" (on a lathe), and can readily be cut and drilled with both hand and power tools.	Weight: Medium Hardness: Low Strength: Medium Working: Medium
California laurel	Umbellularia californica	Western U.S.	A good middle-of-the-road hardwood: easily works with tools, good weight-to-strength ratio.	Weight: Medium Hardness: Low Strength: High Working: Medium

Table 3-1 Hardwoods (Continued)

Common Name	Latin Name	Source	Description	Qualities
Coachwood	Ceratopetalum apetalem	Australia	Coachwood is commonly used for fine furniture; its high strength-to-weight ratio and ease of working make it ideal for even small robot parts.	Weight: Medium Hardness: Medium Strength: High Working: Medium
Utile	Entandro-phragma utile	Africa	Utile is a mahogany-like wood that has a medium density but very high strength. It's reasonably easy to work.	Weight: Medium Hardness: High Strength: High Working: Medium

One application for low-strength woods, oddly enough, is as a reinforcement or as filler. One common application for balsa, for example, is as a shim or spacer.

- **Low-strength hardwoods** Alder, balsa, basswood, butternut, chestnut, cottonwood, elm (American and European), poplar, sassafras, sycamore, willow
- **Low-strength softwoods** Douglas fir, hemlock, pine, spruce, redwood

Plywoods for Robot Building You've heard that "two light coats are better than one" when painting. Wood is similar. Two thin slices of wood, sandwiched together, can be stronger than the equivalent thickness of a single slice. The reason: Wood is at its weakest across the grain. The grains of the two slices of sandwiched wood are placed at right angles, each slice strengthening the other.

This is the concept behind plywood, a familiar product to most everyone. Plywoods are commonly used in residential and commercial construction for everything from floors to doors. The typical construction plywood is made of softwoods, or a combination of softwoods and hardwoods. The plywood tends to be rather thick—3/8" and thicker is common. As such, they're not wholly suited for use in small robots.

A better choice is specialty plywood for boat building, building model airplanes, and making crafts. A typical aircraft plywood uses birch and consists of 3 to 24 plys—the more plys, the thicker the wood. Typical thicknesses are described in Table 3-2.

Sheet sizes vary depending on the manufacturer and source. Hobby stores commonly carry plywoods in 12" × 12" squares; mail order and specialty wood outlets sell sheets up to 48" × 96". The cost per square foot drops with the larger sheets. If you're buying mail order, remember the extra trucking costs for oversize items.

Medium-Density Fiberboard

You may know *medium-density fiberboard* (MDF) by its trade names Masonite, AllGreen, or Synergite. MDF is a manufactured product, usually available in 4 × 8 foot panel sheets, and is made by compressing wood fibers bound with a resin. The thickness and strength of an MDF panel can be controlled during the manufacturing process. There are quite dense MDF panels that are extremely hard and durable. Others (like particle board) consist of larger fibers and wood chunks and aren't as strong.

MDF panels can be made using either hardwoods, softwoods, or both—pine is common. An environmental advantage of MDF is that much of it is made from so-called sawmill discards; even the sawdust itself can be collected and pressed into usable MDF panels. Various brands and styles of MDF panels are routinely

Table 3-2 Hardwood plys

Metric	Inch	Plys
2.0 mm	0.787"	3 or 5
2.5 mm	0.984"	5
3.0 mm	0.181" *	7
4.0 mm	0.157"	8
6.0 mm	0.236" *	12
8.0 mm	0.315"	16
12.0 mm	0.472" *	24

*3.0 mm = approximately 1/8"
6.0 mm = approximately 1/4"
12.0 mm = approximately 1/2"

available at home improvement stores and lumber yards. Look for cut pieces so that you don't have to purchase an entire 4 × 8 foot sheet. Or find a low-cost paper clipboard at the office supply store. Many use MDF in 1/8" thickness.

The downside to MDF panels, especially the less dense particle boards, is that they are quite heavy for their size. A 12" × 12" sheet of 1/4" MDF can easily weigh several pounds. Other disadvantages include the use of urea formaldehyde or other harmful chemicals that can leach out of the wood over time, the corners and edges can break off more readily than with ordinary wood, and dimensional creep, where the wood can warp when under constant weight.

Common Lumber Dimensions

Now we begin to discuss plank lumber. This wood is milled in common dimensions. Unless otherwise specified, the finished milled dimension is less than the stated size because of saw kerf. When you buy a 2" × 4" piece of lumber, for example, you're really only getting 1 1/2" by 3 1/2" (see Table 3-3).

Plank woods are commonly sold either by the linear foot or by the board foot. Linear feet are simple to calculate; simply measure the length of the wood from end to end.

Board feet measurements consider the height, width, and depth of the wood. One board foot equals 144 cubic inches of material. To calculate board feet, multiply the width in inches by the thickness in inches, and then by the length in inches. Finally, divide by 144. Table 3-4 is a handy table for common 1" thick wood.

Table 3-3 Common milled dimensions

Given Dimension	Actual Thickness	Actual Width
1 × 4	3/4"	3 1/2"
2 × 4	1 1/2"	3 1/2"
1 × 6	3/4"	5 1/2"
2 × 6	1 1/2"	5 1/2"
1 × 8	3/4"	7 1/4"
1 × 10	3/4"	9 1/4"
1 × 12	3/4"	11 1/4"

Table 3-4 Board feet dimensions for 1"-thick lumber

Width in Inches	Length in Feet					
	6	8	10	12	14	16
	Board Feet					
2	1.0	1.33	1.67	2.0	2.33	2.67
3	1.5	2.0	2.5	3.0	3.5	4.0
4	2.0	2.67	3.33	4.0	4.67	5.33
6	3.0	4.0	5.0	6.0	7.0	8.0
8	4.0	5.33	6.67	8.0	9.33	10.67
10	5.0	6.67	8.33	10.0	11.67	13.33
12	6.0	8.0	10.0	12.0	14.0	16.0

Rubber

Rubber is not often used as a structural component in robots but is commonly found in wheels, bumpers, padding, and other applications. There are both natural and synthetic rubbers. A common trait of all such materials is that they are *polymers*—polymers are made of long chains of molecules called monomers. This physical characteristic is shared with plastics, but the manufacture of rubber is different. The polymer links in rubber are created through a process known as *vulcanization*. The vulcanization process uses heat to cure the rubber and form its final polymer chain structure.

What sets rubber apart from plastic is that it is elastic at normal room temperatures (some rubber is made to remain elastic at hotter or colder temperatures). Rubber can be used to produce a variety of pliable materials. Its ability to elongate in one or more directions makes it ideal for use as material for gaskets, springs, tensioners, and shock absorbers.

Natural rubber is obtained from latex, the milky secretion of various plants. A primary source is the tree *Hevea brasiliensis*, which produces Pará (or natural) rubber. Another natural source is the shrub *Parthenium argentatum*. Rubber is softened for processing and is mixed with varying amounts of fillers, pigments (such as carbon black), stiffeners, plasticizers, and other materials—depending on the type of desired properties of the rubber.

Most synthetic rubbers are produced with the by-products of oil and coal refinery. Many synthetic rubbers are copolymers, which means they are made up of several polymer molecules. By varying the ratios of these polymers, it's possible to change the physical properties of the rubber. As a result, countless synthetic rubber materials are on the market today with a wide variety of traits.

Among the earliest synthetic rubbers was styrene-butadiene copolymers, also referred to as Buna S and SBR (the "R" being rubber), and commonly found in automobile tires. Buna N, or NBR (which stands for Butadiene-acrylonitrile copolymer rubber), are copolymers of acrylonitrile and butadiene. These are commonly found in flexible couplings, hoses, and rubber o-rings. Because o-rings come in many different thicknesses and diameters, they can be readily used in robotics as a wheel tread, like that shown in Figure 3-2.

Yet another popular and easily obtainable synthetic rubber is Neoprene (polychloroprene), a heavy-duty material that can also withstand high temperatures. Neoprene is commonly used in food processing where high temperatures—up to the boiling point of water—might otherwise melt or degrade ordinary rubber.

Spandex, also known by its technical term urethane elastomer, is made by combining urethane blocks with polyether or polyester blocks, giving the material a unique blend of strength, hardness, and resistance to oils and many other chemicals. Urethane elastomers have largely replaced the classic rubber in such applications as specialty wheels, seat cushion foam, and of course the elastic material in clothing.

Silicone rubber begins it's life as an organic compound but is really a synthetic blended with inorganic polymers to produce the material commonly found in surgical gloves, wire and cable insulation, and tubing. Silicon RTV is a family of silicone-based rubbers that cure (vulcanize) at room temperature—RTV stands for *room-temperature vulcanizing*. This material is often used as a sealant because it is easy to apply out of a tube and is resistant to mold and fungus. It's also used in producing flexible molds and as an adhesive.

Figure 3-2 O-rings, made of Buna N rubber, are well suited as rubber treads for robotic wheels.

In amateur robotics, rubber can be used as manufactured (like o-rings) or molded to produce wires, bumpers, and other shapes. Rubber that requires vulcanizing at high temperatures needs a metal or glass mold. After the mold and the unvulcanized rubber have been heated, the rubber compound is poured into the mold, which is then allowed to cool. The mold is then opened (if it's a two-piece mold), or the rubber is peeled away from the mold and allowed to cool the rest of the way.

Room-temperature vulcanizing rubber is by far the most commonly used in hobby circles. No heat is needed—though some practitioners of the art like to heat the mold or the liquid rubber to assist in curing. The liquid rubber is poured into the mold, where it sets up and cures. All kinds of rubber parts can be constructed this way. Chapter 10, "Going Further," discusses some common mold-making techniques.

Polymer Clay

One wouldn't normally think of clay as a suitable base material for robotics, but in fact, clay has many potential uses in the robot-making art. Clay can be formed into most any shape imaginable and, when hardened, retain that shape.

You can make clay parts for your robots, but a more common application is to use clay for making molds and models. Clay is easily worked into complex shapes, and it holds patterns and textures well. After firing, the clay mold can be used to produce parts out of various materials, including high-temperature wax, fiberglass, and plastics.

The first clays were organic compounds made of silicate and other earthly minerals. These are still used today. Mixed with the right amounts of water, the clay can be shaped into bowls, dishes, and other implements. When heated, or *fired*, moisture is driven out of the clay and the material becomes hard.

The introduction of polymer clay in Germany in the 1930s ushered in a new era of working with clays. Polymer clay is really plastic (typically *polyvinyl chloride* [PVC], the same stuff used in irrigation pipes), ground into miniscule bits, and mixed with a plasticizer. Like natural clays, polymer clay is shaped while it's still in its flexible state. The clay is heated at a moderate temperature (about 350 to 375° Fahrenheit) for 20 to 30 minutes. This fires the clay and produces a rock-solid piece.

You can purchase polymer clay at any arts and crafts store. Instructions (working times, firing temperatures, and so on) aren't always included with the individual packages of clay. If you're new to polymer clay, purchase a starter kit, which will include several small chunks of clay, some simple clay-working tools, and instructions.

Polymer clay is available in a rainbow of colors, but, when used as a mold-making material, the color is not important. You can use various colors if you plan on using clay to make parts for your robots—perhaps small ornamental pieces, bumpers, head or body shapes (common for android and humanoid robots), or even the basic platform that holds the motors and battery.

Bear in mind that although fired polymer clay is quite resilient, it is breakable. That said, you can purchase clays that are not as prone to breakage—these are recommended. If firing clay to make robot parts, avoid any application that exposes the robot to sudden impacts. A clay-based combat robot is probably not a very good idea!

Plastics

The plastics family is *huge* and encompasses thousands of materials, both natural and synthetic. Many natural plastics tend to be flammable, however (especially those made from cotton fibers), so, for the last several decades, the trend has been to use synthetic plastics that resist fire. Of course, all plastics will melt and burn if the temperature is high enough, giving off toxic gasses in the process. This is the single major disadvantage to using plastics.

All plastics are *polymers*, made by joining individual monomer molecules into longer chains (see Figure 3-3). The links between these molecules largely determine the elasticity of the plastic. Tight, closely fitting links produce a rigid plastic; loose and weak links produce a more flexible plastic.

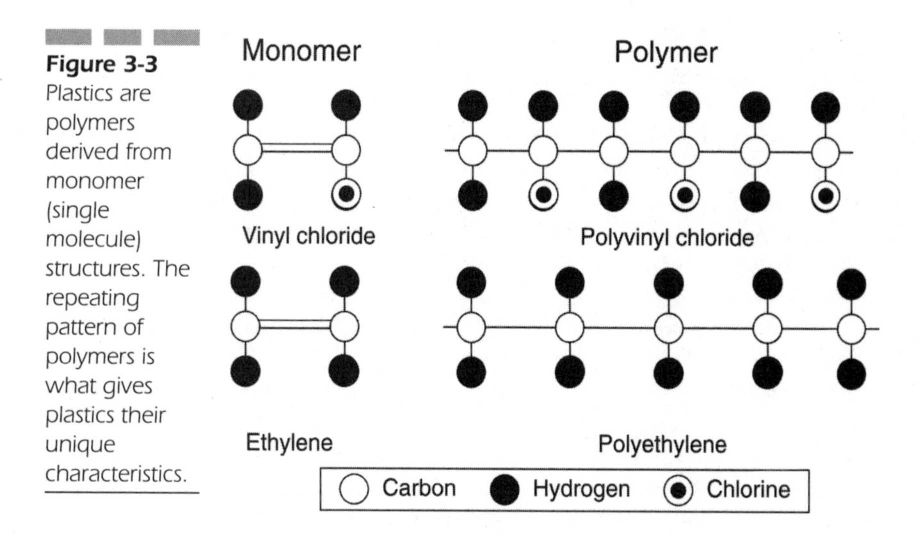

Figure 3-3 Plastics are polymers derived from monomer (single molecule) structures. The repeating pattern of polymers is what gives plastics their unique characteristics.

Plastic materials can be divided into two formal groups—thermosetting plastics and thermoplastics:

- *Thermosetting plastics* develop their polymer-bond structure when heated to a given working temperature. The material is molded or shaped into the desired form at this time. When cooled, the polymer strands join in a complex three-dimensional network, giving thermosetting plastics a characteristic rigidness. Once cooled, a thermosetting plastic cannot be remelted and reshaped.
- Conversely, the polymer bond structure of *thermoplastics* is typically two-dimensional, allowing such materials to take on a variety of physical characteristics from a stiff solid to a gooey liquid, even at room temperature. Thermoplastics are cast and formed by heating them to a working temperature. They can be recast and reformed by heating them up again. Theoretically, thermoplastic material can be heated and cooled in this fashion many times, but in practice, after a number of heating and cooling cycles, the material begins to degrade and weaken.

For consumers, thermosetting plastics are purchased in the form in which they are intended to be used. A good example is Formica tabletop and cabinet laminate. This colored laminate is a thermosetting plastic shaped during its initial manufacture. Conversely, thermoplastics can be reengineered as desired. An acrylic sheet, for instance, can be bent using a heating strip or hot air gun. When cooled, the plastic retains the shape it was molded to.

Injection molding and *extrusion* are primary applications of thermoplastics. Examples include PVC pipe, which is extruded out of an orifice at high pressures to produce its pipe shape, and toy action figures, which are produced by molding various body shapes.

Styrene is a common plastic used in injection molding (see Figure 3-4). The process requires a heavy-duty steel die that is milled with high precision to create the injection mold. The mold is made of at least two parts. The typical arrangement is a front and back mold that join at the sides. In this way, any seams produced into the final figure are not as visible.

The plastic is heated to a liquid state and forced (under great pressure) into the mold. Hydraulic rams keep the mold pieces together until the plastic cools. Once cooled enough that the plastic is no longer runny, the mold opens and the piece is ejected, and then it cools the rest of the way.

Injection molding isn't something most amateur robot makers will undertake, but a similar technique can be employed with lower-temperature materials, such as hot-melt glue. Hot-melt glue is a thermoplastic designed to melt at relatively low temperatures. Given a mold that is small enough, the molten glue can be squirted or poured into the mold, and gravity takes care of the rest.

A benefit of this technique is that the mold does not require preheating, though it will get hot when the melted glue is poured in. Metal and glass are suitable molds for hot-melt casting, but both require long cooling-down periods.

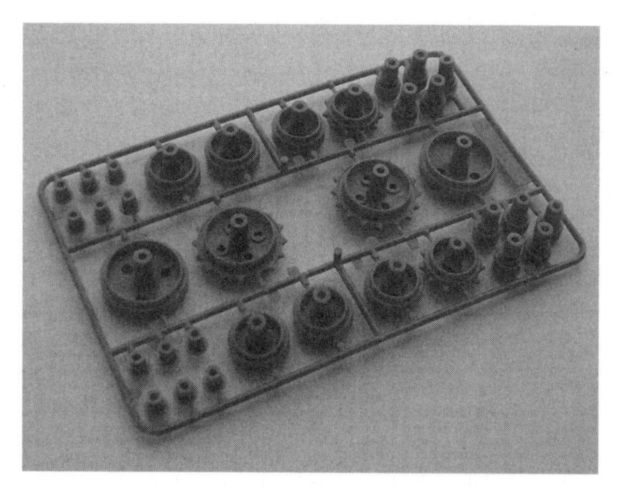

Figure 3-4 Styrene is commonly used to produce injection-molded plastic model kits. Though it is fairly breakable, styrene is easy to mold into any shape.

When a suitable mold-release compound is used, plaster molds are ideal for hot-melt casting, as well as for many common plastics, which have higher melting points than hot-melt glue.

Types of Plastics

Let's forget the 2,000 or so specialty plastics currently available and concentrate on the more common varieties you will encounter as you build robots:

- **ABS (acrylonitrile butadiene styrene)** The most common nonrobot application for ABS is sewer and waste-water plumbing systems. ABS is used in the large black pipes and fittings you see in the hardware store. Despite its shiny black appearance in plumbing material, ABS is really a glossy, translucent plastic that can take on almost any color and texture. It is tough and hard and yet relatively easy to cut and drill. Besides plumbing fittings, ABS comes in rods, sheets, and pipes, and is the plastic used to construct LEGO pieces.
- **Acetal resin** This material is among a group called engineering plastics because it is designed to be used for various engineering applications, such as making prototypes. Acetal resin comes in sheet and block form and can be injection molded. The sheets and blocks are typically cut to shape using computer-controlled mills and routers. The material is known for its dimensional stability—it keeps its shape even under stress.
- **Acrylic** Acrylic is clear and strong, the mainstay of the decorative plastics industry. It can be easily scratched, but if the scratches aren't too deep, they can be rubbed out. Acrylic is somewhat tough to cut without cracking and

requires careful drilling. The material mostly comes in sheets but is also available in extruded tubing and rods.

- **Cellulosics** Lightweight, flimsy, but surprisingly resilient, cellulosic plastics are often used as sheet coverings. They have minor uses in robotics. One useful application, however, is that the material softens at low heat and can be slowly formed around an object. It comes in sheet or film form.
- **Epoxy** Very durable, clear plastic, often used as the binder in fiberglass, epoxies most often come in liquid form and are for pouring over something or onto a fiberglass base. The dried material can be cut, drilled, and sanded.
- **Nylon** Tough, slippery, self-lubricating stuff most often used as a substitute for twine, nylon also comes in rods and sheets from plastics distributors. Nylon is flexible, which makes it moderately hard to cut.
- **Phenolic** One of the first plastics introduced to modern society, phenolics are usually black or brown in color, moderately easy to cut and drill, and terrible smelling when heated. The material is usually reinforced with wood or cotton bits, or laminated with paper or cloth. Even with these additives, phenolic plastics are breakable but quite durable. Phenolic plastic comes in rods, sheets, and pour-on coatings. It has minor application in robotics except as circuit-board material.
- **Polycarbonate** Polycarbonate plastic is a cousin to acrylic but more durable and far more resistant to breakage. Polycarbonate plastics are slightly cloudy in appearance and can be scratched. They come in rods, sheets, and tubing. A common inexpensive window-glazing material, polycarbonates are hard to cut and drill. Lexan is a familiar trade name of a line of polycarbonate materials. Note that although polycarbonate plastics are resistant to breakage, only certain special formulations are bulletproof, no matter what you may read. Don't go shooting your robot as a test!
- **Polyethylene** Polyethylene is lightweight, translucent, and often used to make flexible tubing. It also comes in rod, film, sheet, and pipe form. The material can be reformed with the application of low heat and, when in tube form, can be cut with a knife.
- **Polypropylene** Like polyethylene, but harder and more resistant to heat.
- **Polystyrene** A mainstay in the toy industry, this plastic is hard, clear (can be colored with dyes), and cheap. Although often labeled as "high-impact" plastic, polystyrene is brittle and susceptible to damage by low heat and sunlight. It is available in rods, sheets, and foam board. It is also moderately hard to cut and drill without cracking and breaking.
- **Polyurethane** These days polyurethane is most often used as insulation material, but it's also available in rod and sheet forms. The plastic is durable, flexible, and relatively easy to cut and drill. Polyurethane is also used in the mold-making art.

- **PVC** PVC is an extremely versatile plastic, best known as the material used in freshwater plumbing and outdoor plastic patio furniture. Usually processed with white pigment, PVC is actually clear and softens in relatively low heat. PVC is easy to cut and drill and is almost impervious to breakage. Beside plumbing fixtures and pipes, PVC is supplied in film, sheet, rod, tubing, and even nut and bolt form. Its often overlooked traits are that PVC has poor dielectric properties and does not generate static electricity as much as acrylic or polycarbonate plastics.
- **Silicone** A large family of plastics in its own right. Because of their elasticity, silicone plastics are most often used in molding compounds. Silicone is slippery to the touch and comes in resin form for pouring.

A Closer Look at Thermoplastics

Thermoplastics represent the largest and most useful category of plastics for robotics construction. Three common (but not all-inclusive) families of thermoplastics are cellulosics, polyamides, and vinyls.

Cellulosics

- CAB (cellulose acetate butyrate)
- Cellulose acetate (Rayon)
- Cellulose nitrate (a.k.a. Celluloid—caution: this stuff is flammable!)

Polyamides

- Nylon

Vinyls

- ABS
- Polyethylene (high and low density)
- Polypropylene
- Polystyrene
- PTFE or Teflon (polytetraflouroethylene)
- PVC—flexible
- PVC—rigid

The following is information about several types of thermoplastics:

ABS

- A co-polymer of styrene, butadiene, and acrilonitrile
- One of a number of engineering plastics because of its machinability, yet is also tough and impact resistant
- SAN (styrene acrylonitrile) is similar to ABS and exhibits greater transparency; commonly used for tableware and light fixtures
- Typical commercial uses: tough and impact resistant, commonly used in water-waste pipes and pipe fittings, electronic housings, auto parts, and automotive trim
- Best robotic uses: base material, small parts cut to size and shape
- Workability: easy to drill and cut

Acetal

- Considered an engineering plastic, it is commonly used as a machining material to create pump and value components, gears, and bearings
- Family includes Delrin and glass-filled Delrin acetal copolymers
- High dimensional stability; resists warping due to stress and weight
- Typical commercial uses: parts machining
- Best robotic uses: parts machining
- Workability: designed to be machined; modest difficulty of drilling and cutting

Acrylic

- Chemically known as *polymethyl methacrylate* (PMMA)
- Perspex (U.K.) and Plexiglas (U.S.) are common trade names
- Very good at transmitting light
- Exhibits good internal reflection and can be used as a light pipe; acrylic is often found in lamps and pieces of modern art
- Should be avoided in any application where repeated impact or stress might cause breakage
- Can produce copious amounts of static electricity
- Typical commercial uses: display materials, works of art
- Best robot uses: base material, small parts cut to size and shape
- Workability: difficult to cut and drill without cracking

Cellulosics

- Earliest thermoplastic
- Reasonably tough and impact resistant
- Cellulose acetate in fiber form is Rayon, which is used in some modern motion picture film (polyester is the most common today)
- Available in the forms of acetate and butyrate
- Cellulose nitrate, used in older motion picture film, is highly flammable; seldom used today
- Typical commercial uses: shipping tubes, printing applications, laminations
- Best robotic uses: protective coverings
- Workability: Easily cracked when drilled and cut

Nylon (polyamide)

- At one time, was a trade name, but is now considered a generic term for any of a large family of polyamide plastics, which exhibit high strength-to-weight ratios
- Several types of nylon, distinguished by a number, which is the number of carbon atoms in the reactants used to produce the material (such as 6/6 or 6/12)
- Used as a filament for clothing, rope, and other materials; also used in larger forms as an engineering plastic for producing such things as gears and fasteners
- Very hard and tough
- Low coefficient of friction makes it slippery
- Typical commercial uses: low-friction parts, fasteners, power transmission components (gears)
- Very high dimensional stability (resists warping), though nylon is quite flexible
- Best robotic uses: same as typical commercial uses, especially lightweight fasteners (nuts, screws, and so on)
- Workability: easily machined, easy to drill and cut

Polycarbonate

- Chemical name: poly bisphenol-A carbonate
- Optically clear with some cloudiness common, with very high tensile strength

- Can be combined with glass (20 percent) for added strength
- Best known by the trade names Lexan, Hyzod, and Tuffak
- Typical commercial uses: unbreakable window replacement, machine guards
- Best robotic uses: base material
- Workability: difficult to cut and drill, but not as prone to cracking as acrylics

Polyethylene

- Also known as polythene (mostly in the U.K., as in the Beatles song "Polythene Pam"), but real chemical name is polyethylene
- In raw state is rigid, whitish, translucent, but when processed is flexible
- *High-density* (HDPE) variation—stiffer and higher softening temperature; more transparent. Often used to make kitchen cutting boards; this material makes a strong (though fairly heavy) robot base
- *Low-density* (LDPE) variation—greasy texture and less chemically inert; easily worked at lower temperatures; used commonly in flexible films
- *Ultra-high molecular weight* (UHMW) variation—slippery and impact-resistant; substitute for Teflon (see PTFE)
- Typical commercial uses: protective sheets, trays, cutting boards, and containers
- Best robotic uses: base material
- Workability: low-density materials easy to cut with a sharp knife (or saw when using thicker material), and easy to drill; high-density materials can be hard to saw

Polypropylene

- Like HDPE, but polypropylene has higher surface hardness and is more glossy
- Very low density (lowest among common thermoplastics) so that more goods can be produced for a given weight of material
- Melting point higher than polyethylene
- Can be flexed without fatigue due to its molecular orientation (often used to make integral hinges)
- Typical commercial uses: protective sheets, gaskets, and sporting goods
- Best robotic uses: base material, electrical insulator, flexible joints for arms, legs, or touch bumpers
- Workability: easy to cut with sharp knife or saw, and easy to drill

PTFE

- Stands for *polytetrafluorethylene*, which represents a large family of fluoropolymers
- Expensive to produce, though reprocessed PTFE is available at reasonable prices
- Low coefficient of friction; approximates ice
- Withstands relatively high temperatures, has good impact strength, and is easy to machine
- Materials include Teflon (PTFE), Tefzel (ETFE), FEP, PFA, Kynar (PVDF) and CTFE
- Typical commercial uses: specialty materials
- Best robot uses: custom-made, low-friction components, such as skids
- Workability: machinable and easy to cut and drill

Polystyrene

- Polystyrene is commonly provided in sheet form for thermoforming (vacuum forming); also in granular form
- Cheap and easy to produce
- Commonly used in injection molding for inexpensive goods
- Tends to be quite brittle, though is easy to machine
- Variation: High-impact polystyrene is slightly less brittle, and is found in toys and plastic models
- Typical commercial uses: food containers and injected molded parts
- Best robot uses: small, lightweight parts that are not under stress or weight
- Workability: easy to drill and cut, though breakage can occur

PVC

- Stands for polyvinyl chloride
- One of the most common plastics used today
- Always used with varying amounts of plasticizer; the more plasticizer that is added, the softer the material
- Material is produced as extrusions, films, granules, and pastes (plastisols)
- Can be combined with other plastics for multiproperty benefits (example: ABS/PVC)
- Comes in expanded (or foamed) form; light and strong

- Typical commercial uses: everything, even your credit cards
- Best robot uses: base material, small parts cut to size and shape
- Workability: easy to drill and cut

Additional specialty thermoplastics include Noryl polysulfone, *polyetherether-ketone* (PEEK), carbon-filled PEEK, glass-filled PEEK, Ultem (polyetherimide), polyimide, Torlon (polyamide-imide), glass-filled polyamide-imide, and others. These tend to be quite expensive and are typically used for high-end manufacturing tasks and precision products.

A Closer Look at Thermosetting Plastics

Recall from earlier in the chapter that once set into shape, thermosetting plastics cannot be reheated to change their shape. Thermosettable plastics are among the first plastic products introduced and they are still common today. The following are common types of thermosetting plastics:

Alkyds

- Family name for polyester
- Used in everything from garments to plastic-casting liquids
- Is cold-setting—does not require heat to cure
- Often used as glass-reinforced plastic structures

Aminos

- Category includes *urea formaldehyde* (UF) and *melamine formaldehyde* (MF) resins; in raw state, UF resins are white and can be colored
- Formica and other countertop laminates are in this family
- Very hard and tough
- Good abrasion resistance

Epoxies

- Epoxy resins are yellowish liquids
- Is cold-setting
- Good adhesion to materials and is often used as a strong glue
- Also used as a plastic-casting medium

Phenolics

- Phenol resins are also known as *phenol formaldehyde* or PF resins
- Invariably dark colored
- Cheap to make
- Very rigid; good impact strength
- Easily molded

Polyurethanes

- Made by combining two components, isocyanates and polyols
- Different characteristics achieved by varying proportions of these components
- Often used to produced foamed materials, in rigid and flexible grades
- High strength-to-weight ratio

Common thermosetting plastics include

- **Bakelite** Provided in molded shapes only, such as electrical outlets, appliance parts, and equipment handles.
- **Fiberglass** Excellent impact and tensile strength; can be purchased in castable form (separate glass matting and resin) or in ready-made forms: sheets, tubes, and so on. Moderately easy to cut and drill.
- **Formica (or similar countertop material)** MF thermosetting plastic is laminated under high pressure to a stiff carrier (a phenolic resin); it can be cut using saws or router bits and is extremely tough.
- **Garolite** Similar to fiberglass, Garolite is formed from fibers or glass fabrics woven in resins to form a laminate. Available in ready-made forms. Modestly easy to cut and drill.
- **Polyester** Optically clear, polyester is used as a lower-cost alternative to polycarbonate. It comes in two general forms: *polyethylene terephthalate* (PET) and *polyethylene terephthalate glycol* (PETG) (a copolymer). Available in thin films, rods, sheets, tubes, and other ready-made forms. Static-dissipative polyester resists static charge and does not attract contaminates. Useful for antistatic surfaces and containers. Polyesters are easy to machine, cut, and drill.

Common Plastics Trade Names

Synthetic plastics are engineered products and, as such, are commonly known by their brand or trade names. Because trade names are used so often in the plastics industry, it can be difficult to know exactly what you're getting. In only a few instances are trade names suggestive of the material used to produce the plastic.

Table 3-5 lists many common plastic trade names, and the plastic name or family they belong to. This list is by no means exhaustive. If you encounter a plastic brand that is unfamiliar to you, refer to the manufacturer's Web page, and look for the *materials safety data sheet* (MSDS), which in most cases will indicate the composition of the material.

Table 3-5 Common plastic trade names

Trade Name	Chemical Name/Family
Absylux	ABS
Acetron	Acetal
Acrylite	Acrylic
Alkathene	Polyethylene
Altuglas	Cast acrylic sheets
Araldite	Epoxy resin
Axxis PC	Polycarbonate sheets
Bakelite	Phenolic molding
Baylon	High-density polyethylene
Biberloc	Glass-fiber-reinforced PVC
Celmar	Polypropylene
Celtec	PCV (expanded)
Corzan	CPVC
Cycolac	ABS
Cycoloy	Polycarbonate/ABS
Darvic	PVC
Delrin	Acetal (POM)
Durethan	Polyimide
Eccofoam	Polyurethane resins
Epikote	Epoxy resin
Epophen	Epoxy resins

Table 3-5 Common plastic trade names (Continued)

Trade Name	Chemical Name/Family
Ferex	Fire-retardant polyurethane foams
Fluon	Polytetrafluoroethylene
Glasflex	Acrylic
Hypalon	Chlorosulphonated polyethylene
Hytrel	Polyester elastomer
Hyzod	Polycarbonate
Kapton	Polyimide
Kevlar	Nylon
Kytec	Fluoropolymer
Komatex, Komacel	PVC (expanded)
Levapren	Ethylenevinyl acetate
Lexan	Polycarbonate
Lucite	Acrylic
Lustran	ABS
Makrolon	Polycarbonate
Mylar	Polyethylene terephthalate film
Neoprene	Polychloroprene rubber
Nordel	Hydrocarbon rubber
Noryl	Polyphenylene oxide
Nylon*	Polyamide
Perspex	Acrylic
Plexiglas	Acrylic
Saran	PVC
Silastic	Silicone mastics
Sintra	PVC (expanded)
Stabar	Polyetheretherketone films

Table 3-5 Common plastic trade names (Continued)

Trade Name	Chemical Name/Family
Styron	Polystyrene resins
Sylgard	Silicone elastomers
Teflon	Polytetrafluoroethylene
Tuffak	Polycarbonate
Torlon	Polyamide-imide
Trovicel	PVC (expanded)
Ultem	Polyetherimide
Valox	Thermoplastic polyester
Vamac	Ethylene acrylic elastomer
Victrex-peek	Polyetheretherketon
Vitec	Fire-retardant polyurethane foams
Viton	Fluorinated elastomer
Zytel	Nylon resin

*Nylon is now used as a generic term.

Note that while a trade name may relate to a particular family of plastics, many such plastics are not considered generic. That is, a given trade name may be for a line of nylon-based products, the plastic may employ various patented manufacturing techniques, or it may offer certain engineering enhancements. A good example is the very first plastic noted in the table, Absylux. This is an ABS plastic, but with a conductive additive to make the plastic antistatic.

Common Plastics Acronyms

Apart from trade names (see Table 3-5), plastics are referred to by their chemical name. But chemical names can be difficult to pronounce and remember, so acronyms are commonly used as a substitute. A good example is PVC, for polyvinyl chloride. Table 3-6 lists even more popular plastic acronyms you may want to memorize (no, there will not be a test).

Table 3-6 Common plastics acronyms

Acronyms	
ABS	Acrylonitrile butadiene styrene copolymer
AN	Acrylonitrile
ASA	Acrylonitrile styrene acrylic ester copolymer
CPE	Chlorinated polyethylene
EAR	Ethyl acrylate rubber
EP	Epoxy resin
EPDM	Ethylene propylene diene monomer
EPR	Ethylene propylene rubber
ETFE	Ethylene tetrafluoroethylene copolymer
EVA	Ethylene vinyl acetate copolymer
FEP	Perfluoroethylene propylene
FRP	Fiberglass-reinforced polyester
FRTP	Fiber-reinforced thermoplastic
HDPE	High-density polyethylene
HIPS	High-impact polystyrene
LDPE	Low-density polyethylene
MF	Melamine formaldehyde resin
PA	Polyamide
PAI	Polyamide-imide
PAr	Polyarylate
PB	Polybutylene (Polybutene)
PBI	Polybenzimidazole
PBT	Polybutyleneterephthalate
PC	Polycarbonate
PCTFE	Polychlorotrifluoroethylene
PE	Polyethylene (Polyethene)

Table 3-6 Common plastics acronyms (Continued)

Acronyms	
PEEK	Polyether ether ketone
PEI	Polyether imide
PES	Polyether sulphone
PET (PETE, PETP, PETG)	Polyethylene terephthalate
PF	Phenol formaldehyde resin
PI	Polyimide
PIR	Polyisocyanurate
PMMA	Polymethyl methacrylate
POM	Polyoxymethylene (Acetal)
PP	Polypropylene
PPE	Polyphenylene ether
PPO	Polyphenylene oxide
PPS	Polyphenylene sulphide
PS	Polystyrene
PSO	Polysulphone
PSU	Polysulphone
PTFE	Polytetrafluoroethylene
PU	Polyurethane
PUR	Polyurethane rubber
PVA	Polyvinyl alcohol
PVAC	Polyvinyl acetate
PVC	Polyvinyl chloride
PVCC	Chlorinated polyvinyl chloride
PVDC	Polyvinylidene chloride
PVDF	Polyvinylidene fluoride
PVF	Polyvinyl fluoride

Table 3-6 Common plastics acronyms (Continued)

Acronyms	
SAN	Styrene acrylonitrile copolymer
SB	Styrene butadiene coplymer
SBR	Styrene butadiene rubber
SI	Silicone
SIR	Silicone rubber
UF	Urea formaldehyde resin
UP	Unsaturated polyester resin

Properties of Common Plastics

There is no perfect plastic. Each variety of plastic enjoys a known set of properties that make it useful for one application or another. Some are soft, supple, and virtually unbreakable; others are hard and rigid, and can snap with hardly any effort. A plastic may be perfectly suited for one application and terrible for another.

Table 3-7 is a summary of plastics practical for use in amateur robotics (and reasonably available through various sources), their properties, and how they are commonly used in nonrobotics applications.

Decoding Plastic ID Symbols

You've seen them on the bottom of bottles and canisters. They're the funny circular arrow thingamajigs with numbers inside. You know this already—these funky graphics are *ID symbols* designed to help consumers and recyclers tell the difference between various plastics.

If you're building robots out of junk you find around the house, it can be useful to know what kind of plastic it is, in case you want to join two or more pieces with some solvent cement. Why is this important? Different plastics require different kinds of solvent cements. A solvent for one plastic may do absolutely nothing for another. Figures 3-5 through 3-11 show what the ID symbols mean.

Table 3-7 Properties of common plastics for robotics

Name	Family	Properties	Commonly Used For
ABS (*acrylonitrile butadienestyrene*)	Thermoplastic	Rigid; very strong, tough, and resistant to moisture and chemicals. Is easily scratched. Available in a wide range of colors.	Waste-water pipe; sheets, bars, and rods for mechanical parts; some plastic toys (LEGO)
Acrylic (polymethyl methacrylate)	Thermoplastic	Rigid, optically clear to opaque, very hard. Readily machined, cemented, and polished, but can scratch easily. Good electrical insulator; can generate copious static electricity when rubbed against nylon and other common household materials.	Sheet and rods for structure; skylights; displays and signage; housewares (colored bowls)
Expanded polystyrene	Thermoplastic	Buoyant, lightweight; an excellent insulator for heat and sound. Absorbs shocks; can degrade (crumble) easily.	Sound insulation; heat-resistant packaging
High-density polyethylene	Thermoplastic	Stiff, hard, and fairly resistant to scratches. Good chemical resistance.	Bottles; milk crates; housewares (such as buckets); machine parts; kitchen cutting boards
Low-density polyethylene	Thermoplastic	Tough and flexible; good chemical resistance. Attracts dust unless made with antistatic additives. Excellent electrical insulator.	Squeeze bottles; toys; packaging film

Table 3-7 Properties of common plastics for robotics (Continued)

Name	Family	Properties	Commonly Used For
Nylon (polyamide)	Thermoplastic	Tough and hard; resistant to creep (deformation due to continued stress). Good bearing surface; self-lubricating. High melting point. Very resilient to wear and friction.	Gear wheels; bearings; fasteners; tool casings; packaging
Polyester resin (unsaturated polyester resin)	Thermoset	Stiff, hard, and brittle; highly exothermic (can lead to cracking), good insulator of electricity and heat.	Plastic garden furniture; boats; car repair material; ducting; castings
Polystyrene (high-impact polystyrene)	Thermoplastic	Easy to drill and cut; readily breaks. Available in wide variety of colors. Join using a liquid polystyrene cement.	Plastic model toys; plastic sheets for vacuum forming
PVC, plasticized (polyvinyl chloride)	Thermoplastic	Soft and flexible, with wide range of colors.	Clothing; sealing compounds; dip coatings; electrical wiring insulation; floor coverings
PVC, rigid (polyvinyl chloride)	Thermoplastic	Hard, stiff and tough at room temperature but has a fairly low melting point. Lightweight. Excellent for use in fabrication. Resistant to most alkaline and acidic chemicals; melts with acetone. Poor electrical insulator; doesn't generate much static as a result.	Irrigation pipes, rain gutters; construction materials (sheet, rod, tube, and so on)

Figure 3-5
PETE—polyethylene terephthalate

Figure 3-6
HDPE—high-density polyethylene

Figure 3-7
PVC—polyvinyl chloride

Figure 3-8
LDPE—low-density polyethylene

Figure 3-9
PP—polypropylene

Figure 3-10
PS—polystyrene

Figure 3-11
A combination of plastics, or none of the above

The Joys of Rigid Expanded PVC

What (sort of) looks like wood, drills like wood, cuts like wood, and sands like wood—but isn't wood? It's rigid expanded PVC. This material (see Figure 3-12), commonly used for both sign-making and construction, is manufactured by mixing a gas with the molten plastic. The plastic is then extruded into various shapes: sheets, rods, tubes, bars, and more. The gas forms tiny microscopic bubbles in the plastic, which make the material bulkier when it cools.

Rigid expanded PVC contains less plastic than ordinary PVC materials. This has two advantages:

- *Less plastic* = less weight. That's important for building robots where added weight makes the battery drain faster.
- *Less plastic* = less density. This makes expanded PVC easier to drill, cut, and mill. If you've ever cut acrylic plastic, you know it chips and breaks easily, and its high density makes using hand tools a real chore. The

Figure 3-12 Rigid expanded (foamed) PVC comes in a variety of thicknesses and colors.

thinner expanded PVC materials can be cut using a knife, the thicker stuff with an ordinary saw blade.

Rigid expanded PVC is often used as a replacement for wood in outdoor applications, typically where moisture and rot are a problem. The material is formed into familiar wood-like shapes, some even have a wood grain.

As robot builders, we're not too interested in the wood-grain materials. We're after the rigid expanded PVC sheets used to make signs—sign-makers refer to this raw material as *substrate*. It's available in a variety of sizes and thicknesses, in a rainbow of colors: blue, red, orange, tan, black, brown, yellow . . . you name it.

Rigid expanded PVC goes by many trade names, such as Sintra, Celtec, Komatex, Trovicel, and Versacel, but it's probably easiest if you just ask for it by its generic expanded PVC or foamed PVC moniker. Sheets are commonly available in any of several millimeter sizes. Here are some of the more common thicknesses:

- 3 mm, or roughly 1/8"
- 6 mm, or roughly 1/4"
- 10 mm, or roughly 13/32"

Table 3-8 details the weight of a 12" × 12" piece of rigid expanded PVC at various panel thicknesses. Comparable weight per square foot for acrylic plastic is provided and all weights are representative, as some brands are lighter or heavier than others.

Not all colors are available in all thicknesses. Thicker pieces are available in basic black or white only, with other colors being special-order items—often requiring large volumes of several 4 × 8-foot sheets.

Table 3-8 Rigid expanded PVC weight table

	Weight (lbs.) /square feet	
Thickness	Expanded PVC	Acrylic
.080 (5/64")	.287	.547
.118 (1/8")	.429	.729
.197 (3/16")	.722	1.09
.236 (1/4")	.858	1.46
.393 (3/8")	1.03	2.19
.500 (1/2")	1.30	2.91

The same cements and adhesives for PVC irrigation pipes can be used with rigid expanded PVC. I like using a fairly thin cement, such as Weld-On 66. It has a medium consistency and is clear. It can be brushed on, or as shown in Figure 3-13, applied using a suitable needle applicator (you can get cement brushes and applicators at the same place you buy the PVC plastic). You can also use a specialty adhesive, such as Foamex contact adhesive, for fast production assembly.

PVC cements bond by partially melting the plastic pieces and are recommended over traditional adhesives, including hot glue and epoxies. For the best results, I recommend roughing up the pieces to be cemented by lightly sanding the surfaces. Everything has to be squeaky clean for a good bond. You can clean PVC using denatured alcohol.

If you're assembling large quantities of PVC parts, consider the investment of a thermoplastic welder, such as the one shown in Figure 3-14. These welders apply concentrated heat that melts the plastic pieces together. Get a model that has several temperature settings because you'll need to experiment with the proper heat level for a strong, consistent weld. Thermoplastic welding kits for everything but heavy-duty assembly are available in the $200 to $500 range. The tips wear out over time and replacements cost $25 to $50.

The Case Against PVC

Plumbers and robot builders love PVC. According to industry sources, it's the second most commonly used plastic in the world. And to some, it's also a major polluter.

The international Greenpeace organization feels that PVC is one of the world's largest sources of dioxin. The dioxin is released during the manufacturing of the plastic, but it also results when the plastic is burned. Unlike many

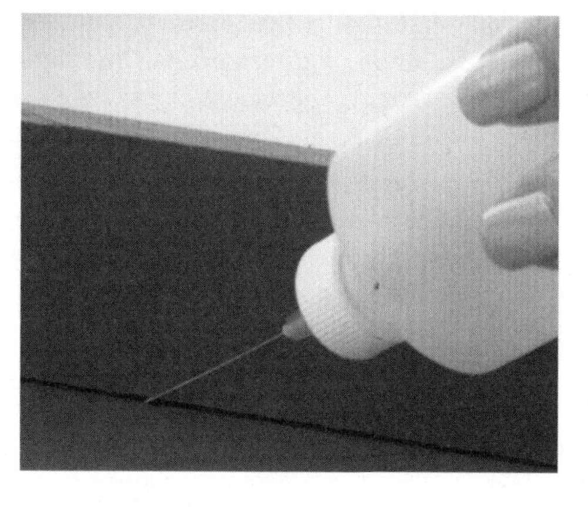

Figure 3-13 A needle applicator can be used to meter precise amounts of water-thin cements.

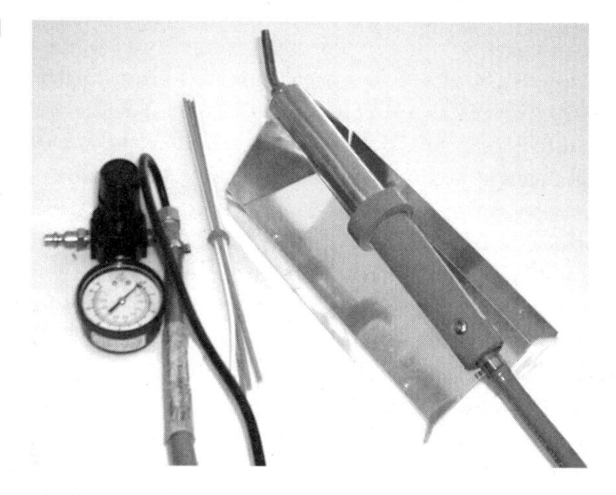

Figure 3-14 This thermoplastic welder welds plastic.

other plastics, most PVC products are not recycled (in the traditional sense) and more often than not are simply disposed of in landfills or incinerated.

Does this mean you should not use PVC in your robotic creations? That's up to you. There are disadvantages to most any substrate or building material you may use. A metal or wood base may be no less environmentally unfriendly; toxins are released during the manufacture of all metals, and the hardwoods in high-quality plywood may come from rain forests that are not being replanted. What's a robot builder to do?

- First and foremost is to use your materials wisely. Don't waste. Not only does this save you money, but it helps reduce the waste going into the landfill. Design your robot so that you can use standard size materials with little or no excess. Except for shavings or very small pieces, save all your discards. You can use them for braces and cross pieces.

- Second, some kinds of PVC materials are more air than PVC. Expanded PVC is made by injecting air or another inert gas into the slurry so that the plastic is bulked up. This reduces the weight of a given piece, and less PVC material is used.
- Finally, when possible, raid the scrap and salvage bins at your local plastics store or sign maker. Just because a sheet of plastic has been previously used for some other application doesn't mean it can't be cut up and used to build your next robot. You'll save some money, and you'll know you're doing your part to clean up the earth.

Metal

Before the modern dependency on plastic, metal was the mainstay of the construction material world. Toys were not made of plastic as they are today, but of tin. Even into the 1960s, cars were made with metal inside and out. Today, a good portion of automobiles are plastic, which not only reduces cost, but also weight.

Although plastic has made numerous inroads, metal is still the preferred material for many applications. It's strong and much of it is cheap. It can be molded by heating and bending, and can be machined to form complex shapes.

Metals are routinely broken into classifications: ferrous and non-ferrous. All *ferrous* metals are made from iron (*Fe*, from which ferrous is derived, is the symbol for iron on the Periodic Table of Elements). Obviously, all metals other than iron are *non-ferrous*. This includes copper, tin, antimony, aluminum, lead, titanium, zinc, and even mercury, which is unique because it melts at room temperature.

Iron is used to make *steel*, which is further classified as high carbon, medium carbon, and low carbon. The amount of carbon added to the iron during the steel manufacturing process controls the *ductility* of the metal (ductility is the ease at which the metal is bent, drilled, and cut). High-carbon steel, containing 0.6 percent to 1.5 percent carbon, is very hard and brittle, and is used to make tools. Parts that require machining are typically made from medium-carbon steel (0.3 percent to 0.6 percent carbon). Low-carbon steel (0.05 percent to 0.3 percent), also called mild steel, is used for wrought iron and decorative work. It is the "blacksmith's metal" and is readily shaped, bent, and cut. Often, when someone speaks of iron, it's actually low-carbon steel.

Stainless steel is a special formulation of steel with a minimum amount of chromium added in. The chromium develops a microscopic film on the metal that helps it resist rust and corrosion. More about stainless steel is covered in the section, "A Closer Look at Stainless Steel."

Aluminum is the most common metal used in robot construction projects, partly because of cost, and partly because it is strong yet lightweight. It's also one of the easier metals to cut and drill, and requires only a modest assortment of tools. The aluminum you buy at the hardware store is actually an alloy; raw aluminum (which is manufactured from bauxite ore) has little commercial value as a finished metal. Rather, the alumina metal is mixed, or *alloyed*, with other metals.

The mixture of the alloy varies depending on the desired properties of the finished aluminum material. Aluminum alloys are identified by number. A common aluminum alloy is 6061, which boasts good machinability (it's not difficult to drill or saw), yet is still lightweight and strong. Other alloys are designed to provide a harder and stronger metal that is able to withstand high heat, but it is also much harder to machine. More about aluminum is covered in the section: "A Closer Look at Aluminum."

Copper is one of the few metals that is used in its pure state, though it's also alloyed with other metals. When combined with zinc, copper makes *brass,* and when combined with tin, copper makes *bronze*. All three are soft metals and are relatively easy to cut and drill.

Copper and its alloys can be readily *annealed*, which is the process of heating the metal to high (but not melting point) temperatures, and then allowing it to cool very slowly. Annealing is used to change such properties as the softness and ductility of the metal, and to rearrange the microscopic grains of the material. It's commonly used to make springs and so-called springy metal.

Zinc and tin are most commonly employed as alloy ingredients or as coatings. In addition to making brass, *zinc* is also used to plate and *galvanize* steel, a dipping process that coats the steel for a relatively rust-free surface. (Galvanizing leaves a tell-tale crystalline look to the surface of the metal.) *Tin* is a malleable bright metal that is commonly used as a coating for steel, as tin resists corrosion and inhibits rust. Tin cans for food are actually made of steel and are then coated with tin to protect against rust. They would be more accurately termed tin-plated steel cans.

Zinc-plated and galvanized steel is common in metal fasteners, as well as various hardware items found at home improvement stores. What is routinely called an angle iron—used to joint two pieces at right angles to one another—is medium-carbon steel that has been galvanized or zinc-plated.

Galvanizing is a critical step in protecting steel (iron) from the ravages of air and electricity. Rust is the natural by-product of the mixture of iron and oxygen. For most structures, even robots, rust is undesirable, and a coating over the steel is necessary to prevent oxygen from reaching the base metal. Paints are often used, but these easily scratch off from the smooth surface of the metal. More about galvanizing can be found in the section, "Benefits of Galvanization."

Other metals, some exotic and some not, also find their way into the robotics workshop. These include nickel, typically used as plating; titanium, for machining very strong parts; iron, for casting and machining; tungsten carbide; cerro alloys, for casting; and lead, also used for casting.

A Closer Look at Aluminum

As previously mentioned, aluminum is an alloyed metal, and the percentages of various metals determine the properties of a given aluminum alloy. Table 3-9 summarizes common aluminum alloys, which are identified by a three- or four-digit number.

In Table 3-9, *ksi* stands for kilograms per square inch (a measure of pressure). The higher the value, the more stress the metal can withstand. *Brinell* is a measure of hardness; the higher the Brinell, the harder the material. At first glance, it may seem that the higher the better, but this is not always the case. A very high hardness is not always desirable when machining aluminum, for instance. Also, a hard metal will wear down the cutting tool faster.

Aluminum alloys are also commonly listed with their *temper* rating. For example, aluminum listed as 6063-T52 is alloy 6063, with temper rating of 52. Tempering is used to change the ductility and other properties of the alloy. The letter designation indicates how the material was tempered:

F—as fabricated

H—strain hardened

T—heat treated

O—annealed

W—solution treated

The first digit of the alloy indicates its chemical makeup as shown in Table 3-10.

Additionally, aluminum-bronze alloys (630, 642, 954, 959) are used for casting or machining replacement parts. You'll find these specialty alloys often used for machining gears, worm drives, and other power transmission components. Other alloys are made for casting rather than machining. Aluminum alloy 319, which has a melting range of 960 to 1200° F, is available in ingot form, and is used to cast parts.

Except for ingots, most aluminum alloys are available in a wide variety of shapes, including plate, sheet, rod, bar, tube, honeycomb, foil roll, discs, and various extrusions—T-shaped, I-shaped, H-shaped, and U-shaped channels, angles, beams, and more. You will find the most basic shapes at the local hardware or home improvement stores; the rest can be found at a well-stocked industrial metals outlet. Check the Yellow Pages for one near you. Additionally, mail order is an option. For example, Online Metals (***www.onlinemetals.com***) offers various aluminum alloys in a variety of shapes.

Aluminum is available as-milled or anodized. *Anodizing* protects the aluminum from corrosion. Anodizing is commonly found on extruded lengths of angle and channel stock at the hardware store and is clear (silver) in color, black, or gold—other colors are also available from metalworking shops that perform

Table 3-9 Properties of common aluminum alloys

Alloy	Machinability	Weldability	Yield Strength	Hardness	Melting Range	Typical Applications
2011	Excellent	Fair	43 ksi	95 Brinell	1005–1190° F	Machinable parts
2017	Good	Good	40 ksi	105 Brinell	955–1185° F	Machinable parts
2024	Fair	Good	47 ksi	120 Brinell	935–1180° F	Fasteners; fittings; wheels
3003	Good	Good	21 ksi	40 Brinell	1190–1210° F	Sheet metal work; equipment tanks
4032	Good	Fair	46 ksi	120 Brinell	990–1060° F	Machinable parts
5052	Good	Good	28 ksi	60 Brinell	1125–1200° F	Drums; tanks
6013	Excellent	Fair	62 ksi	130 Brinell	1052–1195° F	Machinable parts
6061	Fair	Good	40 ksi	95 Brinell	1080–1205° F	Pipe fittings, scaffolding
6262	Excellent	Good	55 ksi	120 Brinell	1080–1205° F	High-temperature engine fittings and parts
7068	Poor	Fair	99 ksi	190 Brinell	890–1175° F	High-temperature engine fittings and parts
7075	Poor	Fair	73 ksi	150 Brinell	890–1175° F	Gear and other power transmission parts

Table 3-10 Chemical makeup of the alloy

Alloy Series	Major Alloying Metal
1xxx	Pure aluminum; for smelting
2xxx	Copper
3xxx	Manganese
4xxx	Silicon
5xxx	Magnesium
6xxx	Silicon and magnesium
7xxx	Zinc

anodizing. As-milled aluminum is available for when the metal is expected to be machined, painted, or anodized by the user.

Though aluminum does not rust, it can corrode if not protected against oxidation. When not anodized, aluminum is usually painted, either with ordinary spray paint, or with *powder coating*. Powder coating is a form of painting where a pigment and a resin are electrostatically charged, then sprayed onto an electrically grounded metal. The result is a uniform coating that is not as susceptible to sagging or dripping, as is the case of aerosol spray painting. Powder coating is primarily the domain of specialty metal coating shops, but there are some consumer-level paint application units you can purchase for small jobs.

A Closer Look at Stainless Steel

So-called stainless steels are special steel alloys that resist corrosion and rust—though they can and do stain. Stainless steel is produced by adding at least 10.5 percent chromium to the iron (some sources insist a minimum of 12 percent chromium; I'll let them argue the point). The chromium creates an invisible surface film—chrome-containing oxide—that resists oxidation. This film makes the material passive and corrosion resistant. There are numerous grades of stainless steel alloys designed to offer a selectable mix of corrosion resistance, working temperature, weldability, hardness, formability, machinability, and strength.

The amount of chromium, carbon, or nickel added to the steel yields different families of stainless steel, each with unique properties, as shown in Table 3-11. If you plan on using stainless steel it's important to know what alloy and grade you're getting, especially if you need to weld pieces together.

Table 3-11 Stainless steel families

Name	Common Grades	Magnetic	Hardening	Welding	Typical Uses
Martensitic	Grades: 410 (most used); 420 (cutlery); 440C (high hardness)	Yes	Yes	Poor	Knife blades, surgical instruments, fasteners, springs, shafts
Ferritic	Grade: 430 (most used); 409 (high temperature)	Yes	No	Poor	Automotive pipes, cooking utensils, building trim
Austenitic (contains nickel)	Grades: 304 (most used), 310 (high temperature), 316 (good corrosion resistance), 317 (best corrosion resistance)	No	No*	Good	Kitchen sinks and other appliances, food processing equipment, eating utensils, fasteners

*Austenitic stainless steel can be hardened by cold working.

In addition to the alloys shown in Table 3-11, there are combination, or duplex, alloys combining the properties of ferritic and austenitic stainless steels, as well as other alloys using molybdenum and other metals.

Stainless steel is most often purchased by specifying its grade or type. The most common—and easiest to get—stainless steel is grade (or type) 304, sometimes referred to as T304. It is available in sheet, plate, rod, tube, bar, and other shapes.

Austenitic stainless steel may also be labeled by its chromium and nickel content. Grade 304 stainless steel is also referred to as 18-8 or 18/8, which indicates 18 percent chromium and 8 percent nickel. For robot building, 18-8 austenitic stainless steels are most commonly found in metal fasteners. These fasteners are generally stronger than their common steel cousins and are nonmagnetic.

For robotics applications—as opposed to something like food processing—the two most critical properties of stainless steel are its ductility and its weldability. *Ductility* determines how easy it is to bend, drill, and cut; *weldability* determines how readily the metal can be welded using an appropriate torch, such as *metal inert gas* (MIG), or *tungsten inert gas* (TIG). Stainless steel grades that are both reasonably ductile as well as weldable include 203, 302, 303, 304, and 316.

Though harder to work with, grades 321 and 347 are also easily welded and very strong.

Heat Treatments for Metal

Most steels are heat treated in order to enhance certain physical properties, such as the following:

- *Hardening* strengthens the metal, and literally makes it harder. The process also makes the metal more brittle. Hardening is accomplished by heating the metal to specific temperatures, and then cooling it rapidly. Tools are commonly made of hardened steel. For mass-produced products, the hardening is added after the general shape of the tool has been formed. Custom-made tools (like metalworking dies or steel molds) are made from hardened tool steel, using very rugged heavy-duty machinery. The degree of hardness is determined by the speed at which the metal is cooled and by its carbon content.
- *Annealing* softens the metal and makes it more workable. Annealing is accomplished by heating the metal and then allowing it to cool very slowly, sometimes over days. During the cooling process, the grains of the metal can be reoriented to endow the material with unique properties.
- *Tempering* removes some of the hardness and brittleness of steel, and in doing so, makes it tougher. Tempering is often employed after hardening or annealing in order to utilize unique properties of the steel. Like hardening, the tempering process involves heating the metal to a specific temperature, and then rapidly cooling.
- *Case hardening* is a coating process for soft steels and allows relatively low-carbon steels, such as wrought iron, to be hardened. Methods of case hardening vary, but one such process involves dipping the heated metal in a case-hardening compound. The metal is then placed in a furnace to be heated. Only the outer shell of the steel is hardened; the inside core is left soft.

Hardening and tempering can be accomplished in the home shop and are helpful if you're making a robot for heavy-duty chores or for combat. The most common method of hardening steel is to heat the metal with an oxy-acetylene torch. The torch is played over the metal for a uniform heat. The metal is heated to a specific color—which represents temperature—and not to a melting point. After the steel has reached this temperature, it is either left to cool in the air or quenched in water, brine, or another liquid.

Alas, I can't provide more details on the ever-exciting topic of hardening metals. There's much more to steelworking than can be discussed in a book like this.

Consult a good metalworking text on the desired temperatures or quenching liquids to use based on the type of steel you are working with and the desired hardness. Pay special attention to the safety section of the book—needless to say, red-hot metal can be a pain to touch!

Benefits of Galvanization

Bonded coatings, such as that provided by galvanizing, are much more scratch resistant than paint. In galvanizing, zinc is metallurgically bonded to steel, providing longer-life protection. Zinc is more electronegative than steel or iron, and therefore the coating provides cathodic or sacrificial protection of the underlying metal. Cathodic action is caused when two dissimilar conductive materials are in contact. The slight electrical difference caused by the two materials causes electrons to migrate from one to the other. The net effect is a kind of corrosion. One of the materials or both are slowly consumed and wither away. With galvanization, it is the zinc coating that will incur the greatest loss of electrons, sacrificing itself over the steel underneath.

If you work with galvanized steel, avoid welding or brazing the material, or at least take special precautions. When heated to very high temperatures, the galvanized coating gives off toxic fumes (from the liberation of zinc), and some suspect these fumes cause certain cancers.

If you are building a steel frame for a robot, you may want to consider constructing it out of welded, uncoated (not galvanized) steel. If you want the protection of zinc plating, take the frame to a metal shop to be hot-dip galvanized. For services near you, try the Yellow Pages under *Galvanizing-Electroplating*, or a subsection under *Metals-Fabricators*.

WireForm Metal Mesh

Not all metals for robot building are rigid and tough. Another metal material that is useful in certain robotic constructions is wire mesh, which is a metal sheet with holes in it. The metal gives the material stiffness, and the mesh allows it to be shaped to almost any contour. It's commonly used to form the shape of an object for sculpting or molding with papier-mâche, but it's also a suitable material for lightweight robot parts.

Metal mesh is not strong enough for the body of a robot, but the metal can be shaped and reinforced to make a protective undercarriage for your robot, a protective shell around electronics or battery compartments, the scoop of a sumo (wrestling) robot, or the semirigid (and therefore semisoft) bumper or skirt material around the 'bot, which protects the robot and its surroundings. Figure 3-15 shows some samples of metal mesh material.

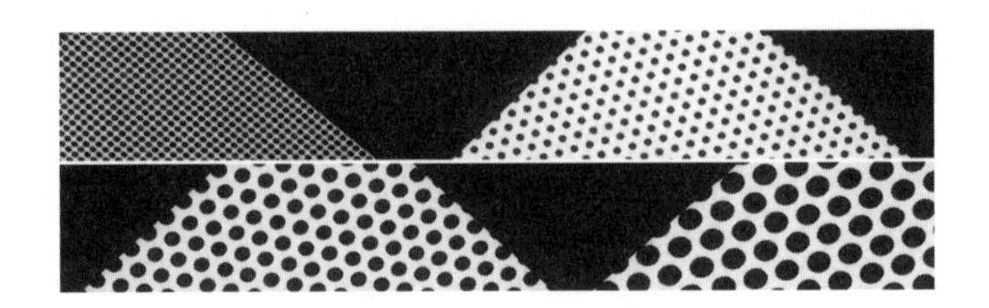

Figure 3-15 Metal mesh can be used to create protective coverings for your robot, as well as scoops, bumpers, and other pieces.

WireForm is a common metal mesh brand. WireForm is a brand made by Amaco; there are others. It is available in 16" × 20" sheets and comes in a variety of meshes and colors. Find it at the better arts and craft stores.

Sources for Metals

You'll find most metals for robot building at the following local sources. If these don't provide the materials you need, try a Google.com search to locate mail-order suppliers of the metals you want:

- *Hardware stores* carry some aluminum and steel sheets, angle brackets, and rods. Selection is limited, and prices can be high.
- *Hobby stores* sell aluminum, brass, and copper in small sheets, rods, tubes, and strips. A common brand sold by stores in North America is K&S Engineering, and you can review this company's product line at ***www.ksmetals.com***.
- *Metal supply shops* that cater to welders are open to the public. Many sell stock in large pieces, which you can have cut so you can get it home in your car. Tip: Check out the remnant bin for odds-and-ends sizes. However, be careful what you buy. If you're looking to weld stainless steel or aluminum, you must be careful to select an alloy that offers good weldability.
- *Restaurant supply stores*, most of which are open to the public, sell many aluminum and stainless materials. Look for spun bowls, cookie and baking sheets, unusually shaped utensils, strainers, and even salt and pepper shakers.

Recap of Metals for Robotics

Table 3-12 reviews metals that are particularly well suited for the construction of robotics. Each metal is noted with its common use, main benefits, and principle drawbacks.

Table 3-12 Metal selection summary

Metal	Common Robotics Applications	Main Benefits	Main Drawbacks
Aluminum	Bases, arms, all structural parts	Reasonably priced; lightweight but strong; easy to cut and drill using proper tools	Plethora of alloys makes picking the right one difficult; can be hard to weld
Brass	Decorative trim, fasteners	Fasteners and trim commonly available at retail stores	Relatively soft; low tensile strength
Bronze	Hose fittings	Readily castable	Fairly heavy
Cerro alloys	Castings	Good alternative to pure lead; low melting points (117–450° F)	Hard to find; can contain toxic metals (such as cadmium, lead, and bismuth)
Copper	Fittings	Ductile and readily machined	Expensive; easily tarnishes
Lead	Castings	Very ductile; low melting point; easy to cast	Toxic
Steel	Heavy-duty frames	Very strong; very inexpensive	Rusts if not protected; can be hard to drill and cut
Stainless steel	Heavy-duty frames	Resists rust and corrosion	Some alloys cannot be welded; can be hard to work
Tin	Sheet metal bodies	Soft and malleable; thin sheets ideal for robot bodies; relatively low melting point (450–725° F)	Can be hard to find in various sizes and stocks
Titanium	Gears, small parts	Extremely hard and strong metal	Expensive; very hard to work without the proper cutting or milling tools; high melting point of 3,000° F

Other Robot Building Materials

Wood, plastic, and metal are common choices for robot building materials, but they are not the only ones available to you. You can build better (and usually cheaper) robots with creative use of common materials. Consider the following.

Construction Foams

So-called construction foam is used in buildings for subflooring, insulation, and sound deadening. This material is commonly referred to as blue foam and represents a rather diverse family of *molded expanded polystyrene* (MEPS) plastics. Blue foam is more rigid than fiberglass insulation, and has a higher *R* factor (an R factor is the rating for insulative materials; the higher the factor, the more insulation the material provides).

Blue foam is available in 1"–3" thicknesses, though specialty versions, such as floor sound-deadening foam, are available in thinner sheets. This latter material can be found at flooring stores.

Blue foam is useful as a substrate or rigidizer for your robot. It weighs very little for its size and bulk, yet offers remarkable rigidity. It's best used when physically cemented to a carrier, such as 1/8" (or even thinner) plywood or plastic. The two materials together provide a strong yet lightweight building platform for your robot. Being foam, it's easy to saw or drill, but it does break off into pieces the way Styrofoam and similar materials do.

Though called blue foam, its color may be either blue or pink. The foam is available in different densities, with many of the pink variety foams having the lowest densities. You may find the heavier blue foam easier to work with because it's not as floppy.

Foam Board

Foam board is part foam and part paper. It is available from art supply stores in thicknesses from 1/8" to over 1/2". Construction is simple: An inner foam sheet is sandwiched on both sides with high-quality paper. Foam board is cheap, is easy to cut with a hobby knife or small scroll saw, and can be readily glued.

You can provide extra rigidity to foam board by combining it with another thick substrate, such as cardboard, thin metal, or plastic. However, depending on thickness and weight loads, foam board can be used by itself.

Foam board is often referred to by a commonly used brand name, Fome-cor. It's available at any art supply store and many crafts outlets.

Art and Sign-Making Substrates

Sign makers use substrates as something to print on. There are many types of substrates for signage, including wood, plastic, metal, foam, and glass. Sign-maker suppliers sell these substrates in convenient sizes and in many colors. Many of the substrates can be used to fashion all kinds of robot parts, including bases, frames, and bodies.

Here's a sampling of sign-making substrates that can also be used in robots.

- **MDO plywood** MDO stands for medium-density overlay, a lighter-weight plywood than stuff at the lumber yard, yet still very strong.
- **Foam board** Foam laminated with plastic, wood, or metal (see the previous section on foam board).
- **Alumalite** Aluminum over corrugated plastic; thicknesses from 1/8"—very strong stuff! Note this is a different material than Alumilite, which is a casting compound.
- **Clad-tex aluminum sign blanks** Variations on an aluminum theme, with plastic, vinyl, and other coatings over an aluminum sheet.
- **Celtec PVC board** Expanded (foamed) PVC—this is the basic stuff you shouldn't do without (other brands include Sintra, Komacel, and Komatex).
- **Coroplast** Inexpensive foam board laminate.
- **Corrugated plastic** Goes by trade names like Gatorboard, Gatorplast, and other Gators; looks like the liner in cardboard but is made of plastic. See Figure 3-16.
- **DiBond** Sheet metal over expanded (foamed) PVC; economical alternative to Alumalite.
- **Econolite** Like Alumalite, but with aluminum on one side only.
- **Fiber-Brite substrates** Fiberglass panels; very light but very strong.
- **PolyCarve substrates** Extruded polyethylene that can be carved into 3D shapes.
- **Reflective tapes** Metal or glass tapes that reflect lots of light.
- **Holographic tapes and films** Add rainbow colors and designs with various patterns.
- **LusterBoard** Aluminum on the outside, lightweight wood on the inside; 1/4" and 1/2" widths common.

Note that prices tend to be very competitive, because sign makers get paid by the job, and they want to reduce their materials costs as much as possible. On the other hand, online retailers of sign-making supplies may require a minimum order of $25 or $50, and some will not cut material to more convenient sizes (typical sign substrate sheets are 4 × 8 feet). Not only do you get a big sheet of

Figure 3-16 Corrugated plastic comes in many colors and thicknesses.

material to contend with, but it must be shipped via motor freight, not UPS. Because of added expenses of trucking, look for sign-maker suppliers that sell already cut pieces.

Cardboard

Common cardboard consists of two thin sheets (liner boards) of paper sandwiched to a corrugated inner core, or medium. The thickness and material used for the inner corrugated medium determines the overall rigidity of the cardboard. Cardboard for shipping boxes is too lightweight for most robot projects.

A better kind is called *honeycomb board*. This is a special kind of cardboard that uses a thick honeycomb medium. Thickness can vary from ½" to several inches. Honeycomb board is available at many art supply stores and is used in model making. For robotics, it can be sandwiched to a thin carrier, such as plywood or plastic, or even metal. The carrier need not be thick; the strength of the sandwich material will come from the cardboard.

Compared to both cardboard and foam core, honeycomb board (that's a generic term for it, by the way) is quite expensive. However, it has good structural characteristics. When used properly, honeycomb board can hold over 100 lbs.

Substrate Sources

Apart from art and craft shops, substrates are readily available from sign-maker supply outlets. You can find them locally or on the Internet. Most will ship materials to you, but be wary of high freight charges. A 4 × 8-foot panel of Gatorboard may weigh only a few pounds, but it could cost $100 or more to ship it to you because of its large size. Look for sources that sell cut pieces, and be sure to get a shipping quote before you order. The following are some places to look for supplies:

Budget Robotics ***www.budgetrobotics.com*** (cut pieces of expanded PVC)

Custom Cut Aluminum ***www.customcut.com***

Fiber-Brite ***www.fiberbrite.com***

Harbor Sales ***www.harborsales.net*** (very nice selection)

Laminator's Inc. ***www.signboards.com***

Lee's Sign Supply, Inc. ***www.leessignsupply.com***

Port Plastics ***www.portplastics.com***

You can find additional sources using Google.com. Here are some search phrases you can try:

"sign-making" substrates

"sign-making substrates"

"alumalite substrates"

"foam core" substrate signs

CHAPTER 4

Selecting and Using Adhesives

Chemical companies would like you to think adhesives are a modern invention. In fact, humans have been using adhesives for thousands of years. True, ancient civilizations didn't have the luxury of a wide selection of glues and cements at their local Home Depot. But they did have natural adhesives like tree sap, food gluten, and insect secretions. It could even be argued that the first Post-It note might have been someone licking the back of a piece of papyrus and sticking it onto the side of a pyramid. Okay, so I said it *could be* argued . . .

Today, we take most adhesives for granted, which is both good and bad. It's good in that a wide variety of adhesives are relatively common and inexpensive. We don't have to give them much thought. It's bad because we don't give them much thought, and this can lead to improper use and frustrating results.

Modern adhesives are chemical concoctions designed to bond two surfaces together, and whether a given adhesive will make them stick depends a great deal on the nature of those surfaces. This chapter reviews the selection and use of a variety of adhesives available to the consumer. These adhesives can be purchased at craft, hobby, hardware, and home improvement stores. The most generic of adhesives can be found even at the grocery store and the discount outlet.

I've intentionally left out most industrial adhesives for three reasons:

- They can be hard to find. By their nature, industrial adhesives are available through distributors and manufacturers, and not at retail stores.
- They come in large quantities. Unless you're George Lucas building an army of synthetic soldiers, you don't need a five-gallon jug of some superadhesive.
- They're more potent than consumer-grade adhesives and, quite often, more hazardous to use. Many industrial adhesives are poisonous and flammable in the extreme, and a few even require specialized training regarding their proper use.

This chapter begins with some general information about adhesives and then discusses the selection and use of several common types.

NOTE: *I haven't made a big deal about the differences between* adhesive, glue, *and* cement. *Some folks claim no difference exists, but in general, I subscribe to the notion that an adhesive (or glue) is a chemical that bonds two materials, without significantly altering either material. A cement, on the other hand, dissolves or otherwise alters one or both joined materials. For the sake of simplicity, I call an adhesive a glue, and vice versa, though some like to define glues as naturally occurring adhesives. As there are very few all-natural adhesives used today, I'll forgo making this distinction.*

Adhesive Setting and Curing

All adhesives bond by going through a number of phases, most notably setting (also called fixturing) and then curing. During the first phase (*setting*), the adhesive transforms from a liquid or paste to a gel or solid. Although the adhesive may look to be hard when set, it is not yet very strong. This requires *curing*. Setting times for most adhesives are measured in minutes or even seconds. But curing takes much longer—typically 12 to 24 hours (see Figure 4-1), and often more.

For *most* adhesives, curing time is greatly dependent on several factors:

- **Air temperature** The warmer the air, the faster the curing (and the faster the setting time, for that matter).
- **Air humidity** Adhesives differ in their affection for moisture in the air. Some, like Super Glue, cure faster when the air is humid. Others, like epoxy, cure faster when it's dry.
- **Surface temperature** Warm surfaces tend to promote faster curing. This is most notable when gluing metal.
- **Adhesive volume** The more adhesive that is applied to a joint, the longer it takes to cure.

The curing process varies by adhesive. As background, here's a short rundown of the most common curing methods of modern adhesives.

- **Chemical reaction curing** Most adhesives cure through some chemical reaction, so this is a generic curing method. The reaction begins via a number of means, such as the mixing of a resin with a hardener, contact with an activator, exposure to air or moisture, or contact with an acidic or alkaline surface (that is, anionic reaction). The exact chemical reaction isn't critical to our understanding of the adhesive in order to use it, but in the interest of a better bond, it helps to know the basics of the process.

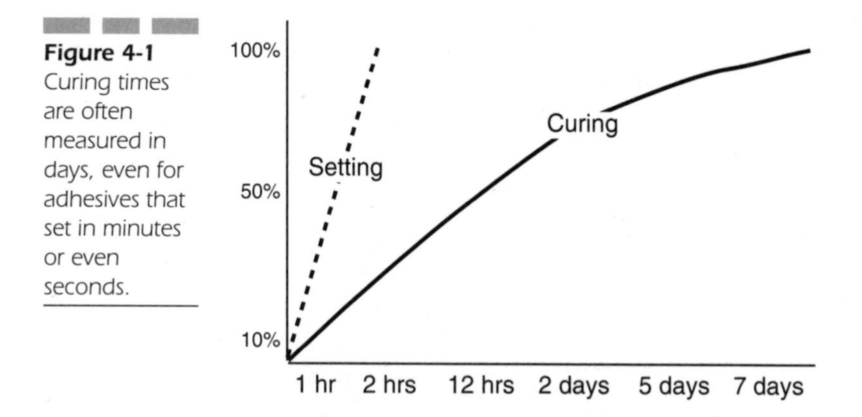

Figure 4-1 Curing times are often measured in days, even for adhesives that set in minutes or even seconds.

- **Catalytic curing** Catalytic curing relies on a chemical reaction when at least one of the components in the reaction is not altered. Note that this isn't the same as solvent cement that may or may not alter the material being bonded.
- **Anaerobic reaction curing** This class of adhesives, commonly used for thread-locking applications because of its sheer strength, cures when it is deprived of oxygen.
- **Pressure sensitive curing** Pressure applied to the adhesive begins a curing process. These adhesives are most common in adhesive tapes.
- **Heat (exothermic) curing** Many adhesives give off heat while curing, and this heat accelerates the curing and makes it complete. Two-part epoxy is a good example of this type of adhesive.
- **Light curing** Specialty adhesives react to ultraviolet or visible light and set when exposed to a bright light source. The brighter the light, the faster the adhesive cures. A common light-sensitive adhesive is used to secure rearview mirrors directly onto windshields of automobiles.

Safety Concerns when Using Adhesives

Like most people, you probably ate the craft paste used in grade school to make paper crafts for mom and dad. Your teacher may have scolded you, but there is little harm in eating the paste, because it's nontoxic, being made of flour or cornstarch, and water.

But because craft paste is harmless doesn't mean other types of adhesives are not. In fact, some are so downright deadly: Ingesting just a little bit can kill you, or at least, make you quite sick. You don't even have to eat it—you can become ill by breathing the vapors. And a few, such as silicone adhesives, are known to contain cancer-causing ingredients.

With this in mind, here are some simple commonsense rules when working with any kind of cement or adhesive:

- Read the label thoroughly before use. This may be obvious, but some folks figure a glue is a glue is a glue. This approach may work for roses, but it's definitely not the case for adhesives.
- Unless you know the adhesive is nontoxic and does not emit any fumes, use only in well-ventilated areas.
- Don't smoke or apply adhesives near an open flame.
- Don't eat while using adhesives.
- Wash up with warm water and soap after handling any adhesive.

- After applying the adhesive, *promptly* recap the bottle or tube. This not only prevents the adhesive from drying out, but it keeps any volatile vapors from escaping.
- When using the really nasty stuff, like most epoxies, download and print the *material safety data sheet* (MSDS) for the product, and keep it handy. These data sheets can be found at the manufacturer's Web site, which these days is usually indicated on the product packaging. You can also try the following Web sites:

  www.msdssearch.com

  www.msdsonline.com
- Keep away from small children (or big children if they have a habit of messing around with potentially dangerous, poisonous adhesives).

General Household Adhesives

I'm grouping a large and admittedly diverse collection of adhesives under the moniker of household, because they are traditionally used for general household bonding chores. These adhesives, sold by Loctite, Elmer's, Weldbond, and dozens of others, are commonly available in discount and department stores, home improvement and hardware stores, online supply outlets, and more. They are an ideal first choice for a wide variety of basic gluing needs.

Binder Types

Household adhesives include the following binder types:

- *Casein* glues are derived from the protein of cow's milk, sodium caseinate. I suspect this is why they use a cow for the Elmer's brand of glues (the cow's name is Elsie, who used to grace other Borden products). However, like many brands of similar products, Elmer's now uses an adhesive based on *polyvinyl acetate* (PVAC) (see the following bullet), and casein glues are primarily the domain of specialty adhesives for woodworking. In the past, similar glues were derived from other organic proteins, such as animal bones and hides, but these are all but obsolete in most areas of the world. A popular application for casein glues includes paper and wood bonding. Casein is water-soluble and can be thinned as needed. Note that there are some recipes on the Internet for using Elmer's for cooking, citing its casein content. These people aren't aware the product changed its ingredients to keep up with the competition. Yuck!

- *PVAC* dispersion adhesives are among the most popular general-purpose glues now available. Most of the white and yellow woodworking glues (Figure 4-2), such as those from Elmer's and Weldbond, are PVAC-based. They are water based, easy to clean up, and inexpensive. Although their bond strength isn't as high as other adhesives, PVAC adhesives can be successfully used to bond a porous material (paper, wood, and so on) to another porous material, or to metal, plastic, or even glass. An advantage of PVAC adhesives is that they don't emit harmful fumes, except for a slight vinegary odor.
- *Styrene-butadiene rubber* (SBR) binders are among the most common of all water-based adhesives, used for both household and automotive applications. The adhesive remains rubbery and is typically used to form gaskets and sealants. This type of adhesive is popular for temporary bonds—when you want to glue something today but be able to take it off tomorrow.
- *Silicone-based* adhesives are used for both gluing and sealing and are used for similar tasks as SBRs. They can be used to bond most any nonporous surface to another. A common trait of silicone adhesives is that they remain elastic, even in cold temperatures. Another all-too-common trait is that many contain *trichloroethylene* (TCA), a toxin and carcinogen. Use only in well-ventilated areas. Most silicone adhesives cure at room temperature (hence their moniker, *RTV*, which stands for *room temperature vulcanizing*) by exposure to the moisture in the air. The best bonds are made when it's not bone dry. It's also critical that the bottle or tube be recapped tightly so that no moisture can enter.
- *Urethane* (or polyurethane) adhesive is an isocyanate polymer plastic often used to bond wood, but it is also able to bond other materials, including many plastics and metals. It is particularly popular when bonding metal to wood. It cures by a catalytic reaction with water, and it releases few toxins (compared to many other adhesives) into the air. This family of adhesives is also referred to as MDI, which stands for diphenylmethane diisocyanate.

Figure 4-2 Household PVAC-based adhesives, such as white woodworking glue, are among the most common for general applications.

- *Contact cement* is based on various volatile organic compounds, such as alphatic napthta, hexane, toluene, methyl chloroform, and acetone. As you might surmise, none of this stuff is healthy for you, and contact cements must always be used in a well-ventilated area—or else you will become very sick from using them. Contact cement is designed for joining smooth porous and nonporous materials. As their name implies, these cements are designed to bond more or less instantly on contact. This is accomplished by applying a thin layer of the cement on one or both surfaces to be joined, and then waiting a minute or so for the cement to partially set up. Applying pressure to the joint aids in a strong bond. Popular contact cements include DAP Weldwood, as well as various brands from 3M, such as C-77.
- *Solvent-based* cements use a chemical that dissolves the material it is bonding. They can be tricky to use because if the solvent isn't precisely matched with the material, nothing happens! Perhaps the most common solvent-based adhesive is used for bonding PVC irrigation pipe. It is available at any hardware and home improvement store that sells PVC pipe. The cement comes in either a liquid-thin or medium consistency; the latter is handy for joining parts that have minor gaps. The same solvent cement can be used with expanded rigid PVC sheets, which can be used to construct robot bases. See Chapter 3, "Robot Building Materials," for more details. Other solvent-based cements are available for other plastics: ABS, polycarbonate, acrylic, styrene, and so on. Testors (and others) makes a solvent cement specially formulated for bonding polystyrene-based plastic models. These can be purchased at hobby and craft stores.

Adhesives are commonly sold in tube or bottle form, but others—notably those from 3M—are meant to be sprayed on. 3M Photo Mount is a popular spray adhesive that is used to bond paper, foam, and other lightweight materials. Because it can be sprayed on, it is useful for bonding large surface areas, such as thin aluminum sheets, wood, plastic, foam, or some other substrate.

Using Household Adhesives

With very few exceptions, household adhesives (those grouped previously, at any rate) use single-part chemistries, so there is nothing to mix. Simply open the tube, can, or jar, and apply the adhesive to the surfaces to be joined. Here are some handy tips for working with adhesives.

- Use all the adhesives sparingly. A common mistake is to think that if a little bit of adhesive will do the job, a lot will do the job even better. In fact, the reverse is true. Use only enough adhesive to apply a thin coat to one or both surfaces to be joined. The more adhesive you use, the longer it will take to cure. And some adhesives won't cure completely if they are applied too thickly.

- For adhesives with a watery consistency, apply with a small brush or cotton-tipped swab. For thicker consistencies, apply directly from the tube or with a wooden toothpick or a manicure (orange) stick. The brush, swab, or other applicator must be clean.
- Speaking of clean, very few adhesives will stick to grease and dirt. Always be sure to clean the surfaces to be joined. For plastic and metal, household-grade rubbing (isopropyl) alcohol is sufficient, but be sure *all* the water content has evaporated before applying the adhesive. Use solvents (acetone, lacquer thinner, and so on) sparingly, and then, only when you know they won't hurt the materials you are bonding. For wood, clean using a damp cloth, and let dry. Sand the surfaces as needed to remove stubborn dirt.
- Avoid moving the glued joint until the adhesive has a chance to set. This takes from a few seconds to several minutes. PVAC adhesives set rather slowly, and the joint may need clamping for 20 to 40 minutes to assure a strong bond. Read the instructions that come with the adhesive to determine if the manufacturer recommends clamping.
- Excess or dripped adhesive should be cleaned promptly. PVAC and casein adhesives can be cleaned up with water. For nonwater-soluble adhesives, use alcohol or acetone *before* the adhesive sets and cures. Test the cleaner before applying it to any material to make sure it won't cause damage. If using a cleaner will harm the material, let the adhesive set and then try to peel it off.

Two-Part Epoxy Adhesives

Two-part epoxy adhesives are thermoset polymers consisting of a *resin* and a *hardener* (sometimes also called a catalyst, but this term is not quite accurate, according to some epoxy experts). Separately, these materials remain in a liquid form. When combined, the hardener reacts with the resin, and the mixture begins to set quickly. During this process, a bond is created as the epoxy liquid fills pores, cracks, and crazes in the surfaces of the materials.

Two-part epoxies are known for three distinct advantages:

- The liquid of the epoxy is rather thick and has good gap-filling qualities. This means you can use it to join surfaces that are not precisely aligned to one another.
- Epoxies set up and bond relatively quickly, in 5 to 30 minutes (however, some epoxies are designed to set up much more slowly). Epoxies can bond to many surfaces, including paper, wood, metal, fiberglass, most plastics, and fabric.
- Once set, epoxies are extremely hard and impervious to moisture, most corrosive chemicals, and oils.

The typical package of epoxy adhesive consists of two tubes or bottles: One is the resin, and the other is the hardener. The tubes are separate in some products, and in others, they are joined as one unit, with a single plunger in the center for accurately metering the resin and hardener. Figure 4-3 shows a typical plunger-style tube applicator.

Note that unlike many resin or hardener products (such as polyester casting plastics), the resin and hardener provided in most consumer-grade epoxies are mixed with a 1:1 ratio. This permits easier metering of the liquids; if the liquids are not mixed with the proper ratio, the setting time can be greatly increased, and the strength of the bond can be significantly reduced.

Still, many industrial epoxies are not mixed with a 1:1 ratio, but instead use perhaps 60 percent resin and 40 percent hardener, or some other ratio. It is important to read the instructions *carefully* before use; otherwise, a messy or poor job could result.

It is typical for resin and hardener to be mixed by volume. Measuring cups are a sticky but workable solution—the cups are usually included in the package. Industrial epoxy systems rely on special applicators that accurately meter out the liquids. Small amounts of consumer or industrial epoxy resin and hardener can be mixed on a piece of paper, as detailed in the next section.

Selecting and Using Two-Part Epoxies

Although most two-part epoxies use similar technology, epoxies are made to best suit certain types of jobs and surfaces. The choice is somewhat limited to the general consumer, who can select between 5-minute and 30-minute epoxy. The latter takes longer to set but, as a general rule, provides a better bond.

Yet there are hundreds of additional epoxy formulations, with qualities specific to brands and products. These specialty types are available through mail order (such as through boat making and repair catalogs), better hardware stores, and direct from manufacturers. It's impossible to list them all, but the following are the major alternatives available in epoxies.

Figure 4-3 A plunger applicator helps assure proper metering of epoxy resin and hardener.

- **Viscosity** This relates to the thickness of the post-mixed (resin with hardener) liquid before it has gelled or set. A more watery epoxy is desired for applications where you must flow the epoxy into the joint to be bonded. Normally, a fairly heavy viscosity is desired, because it makes it easier to do a clean job.
- **Filler** Filler is used to bulk up the resin and make the epoxy better suited for bonding to specific materials. Filler for wood is used with epoxies meant for joining pieces of wood, for instance. J-B Weld brand contains filler designed for bonding to metal.
- **Elastic modulus** The term elastic modulus (also *Young's modulus*) is used to describe the relative rigidity of a cured epoxy joint. For most applications, a low to medium elastic modulus is usually preferred, but you can purchase high-elastic-modulus resins and additives if needed. Special curing temperatures are usually required.
- **Colorant** Colors can be added to the mixed liquid. Some alteration of the setting time and bond strength is inevitable, however, so if possible, avoid the use of colorants.
- **Accelerators** An accelerator speeds up the setting and curing time, and is best used when the epoxy cannot be applied under optimal circumstances, as in cold weather. The accelerator can be a separate liquid added during the mix, or it can be included with a given hardener.
- **UV protection** Most epoxy resins are susceptible to breakdown from ultraviolet light of the sun. Additives and special resins are available if you need your epoxy joints to be exposed to full sunlight.

Note that so-called 5- or 30-minute epoxies are termed as such based on the setting time, not the curing time. It takes 12 to 24 hours for most epoxies to cure to 60 percent to 80 percent; then the remainder takes over a period of a *several weeks* (curing is faster with warmer temperatures). Once cured, the bond achieves its maximum strength. Epoxies cure by exothermic action, meaning heat is generated during the curing process. The faster the epoxy cures, the more heat is given off.

To use a two-part epoxy, it's necessary to first mix the materials together, or else use an applicator that thoroughly premixes them. The latter is not common for consumer use, so we'll concentrate on the manual mixing method.

1. Apply short (1" to 2") but equal-length parallel beads of resin and hardener to a small piece of paper. A 3" x 5" index card is ideal. It's always better to mix too little than too much, and you can always mix in more as you need it.
2. Use a wooden (not metal or plastic) toothpick to stir the liquids together. Mixing must be thorough. You can readily determine that mixing is complete if the hardener is colored. One technique you can use to achieve a good mix: Start by mixing the parallel beads with a zigzag action, and then

scoop the material toward a common center. Stir this center dollop for 15 to 20 seconds. See Figure 4-4 for more detail.

3. Apply the mixed epoxy to one or both surfaces to be joined.
4. Most epoxy joints should be taped or clamped to prevent movement of the joint during the setting time. If the joint moves while the epoxy is setting up, the bond will be greatly weakened.

Unused mixed epoxy *must* be discarded. It cannot be reused, and it cannot be put back into the tube. Doing so will ruin the unused material in the tube. Allow the liquid epoxy to harden on the paper card before throwing it into the trash.

NEW PRODUCT TIP: *A handy and relatively new product is the epoxy packet, a small two-part, tear-open package that contains already metered amounts of resin and hardener (see Figure 4-5). Simply snip the top off the package and squeeze the resin and hardener onto your mixing paper. Stir thoroughly and apply the epoxy. The used packet is thrown away.*

Setting and curing times are accelerated by using heat and minimizing moisture (including air humidity). As a general rule, the setting and curing time is reduced by half for every 18- to 20-degree Fahrenheit increase in temperature. This also means you should avoid using epoxy in cold weather, unless you want to wait the extra time for the adhesive to set.

If bonding to metal, preheat the metal before applying epoxy. For small pieces, you can hold the metal in your hands for a few minutes to warm it up. For a larger piece, put it in an oven set at lowest heat for five minutes. Let the metal cool if it's too hot to touch with your bare hands. You want it warm, not sizzling!

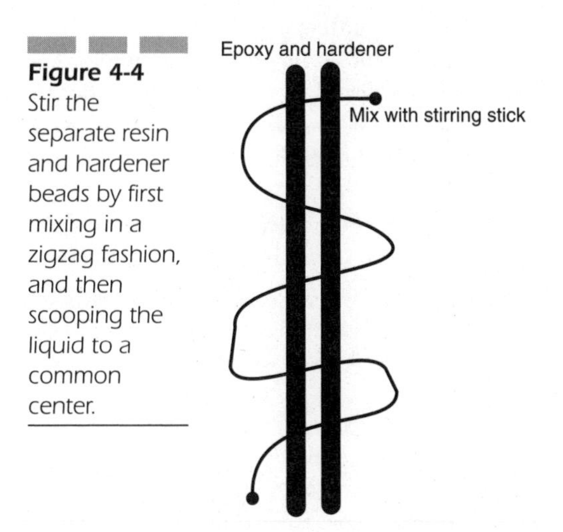

Figure 4-4 Stir the separate resin and hardener beads by first mixing in a zigzag fashion, and then scooping the liquid to a common center.

Figure 4-5 Epoxy packets contain premeasured amounts of resin and hardener and are handy for occasional jobs.

Safety Tips for Epoxies

Both the resin and the hardener used in epoxy are nasty chemicals. Formulations vary, but a common basic resin used in epoxy adhesives and coatings is diglycidyl ether of bisphenol, which is created by a chemical reaction with acetone. The material is flammable and poisonous. Most hardeners are polyamines, chemically similar to ammonia, and are about as irritating to the eyes and mucous membranes.

If you've ever used a two-part epoxy, you know they stink. Well, now you know why. The stink isn't added by the manufacturer for fun; it's part of the chemical makeup of the epoxy, and exposure to these chemicals should be avoided whenever possible. Therefore, always use epoxy in a well-ventilated area, away from open flames. Obviously, you should not smoke or eat while using epoxies and you should always thoroughly wash your hands when the job is done.

Some people are more sensitive to the chemicals in epoxy than others. The vapors from both resin and hardener are known to cause a mild irritation to some, while it makes others violently sick (I am of the latter group). If you get a bad reaction to epoxy, it is best to look for a different adhesive to use. Many others are available.

If mixed epoxy gets on your clothes, you might be able to salvage your duds by wiping the spill with a small cloth dabbed with denatured alcohol or acetone. This must be done before the epoxy hardens. Hopefully, you'll have a change of clothes nearby, because it's best to do this when you're not wearing them. Afterwards, soak the article in water and detergent for 15 to 20 minutes, then wash normally.

Unmixed hardener is alkaline; it can eat through clothing and can cause irritation and even burns on skin. Wash it off immediately if you get any on you.

When measuring out resin and hardener, be sure to replace the covers or caps when done to prevent the volatile fumes from escaping. Not only does this reduce unnecessary toxic vapors from reaching you, but it preserves the resin and hardener, giving both a longer shelf life.

Acrylic Adhesives

Acrylic adhesives are similar to epoxies in that they are composed of two parts, in this case a base adhesive and an *activator*. Acrylic adhesives are known for their strength: They can be every bit as strong as epoxies, if not stronger. The setting time is generally faster than epoxy, however—in some cases, as fast as Super Glue and its competitors (see the following section for more on Super Glue). A strong bond is made almost the moment the two surfaces are brought together.

Although many acrylic adhesive products are available, a typical method for applying them is to coat one surface to be bonded with the activator, and the other with the base adhesive. The bond is made very quickly, so it is important that the surfaces are aligned properly when brought together.

An activator can also be applied to both surfaces when bonding uneven joints. The larger the gap there is in the joint, the longer it takes for the adhesive to set and cure. The adhesive sets very quickly, but takes up to 72 hours (depending on the gap) to fully cure. Once cured, this stuff is like steel.

Other acrylic adhesives are mixed, either manually or with a special plunger application, similar to the way epoxy is mixed. Many better-stocked home improvement stores—particularly those that service contractors—sell both activator-cured and two-part mixed acrylic adhesives. Loctite is a major manufacturer of acrylic adhesives for contractor and industrial use.

Super Glue: The Good, the Bad, and the Sticky

Super Glue is a trade name, but it's also increasingly used as a generic term for a family of adhesives known as *ethyl cyanoacrylates*, or CA. This adhesive is a space-age bonding agent that, thanks to its molecular structure, is able to bond most anything within seconds. CA glues cure upon contact with slightly alkaline surfaces. Moisture on the surfaces that are joined neutralizes a stabilizer in the CA glue so that the surfaces bond to one another. Unlike many other glues and adhesives, CAs do not require heat or other chemicals (an activator or hardener) to cure.

Since CAs were introduced in the 1970s, they've become a mainstay for manufacturing and household repair. They're even used in solving crimes: The fumes from curing ethyl CA compounds are affected by the oil left by fingerprints.

Forensics labs are able to lift prints from many types of surfaces that once hid the evidence of a crime.

Tips and Techniques for Successful Super-Duper Gluing

For all their benefit, CA glues can be finicky. If used incorrectly, they may provide only a weak and temporary bond. Keep the following in mind when using CA adhesives:

- Though some CA glues are available as heavier-bodied gels, the most common CA is water thin. Typical CA glue will not fill voids in the materials being joined. This means the surfaces must be reasonably smooth (a *slight* abrasion of the surfaces may enhance the bond; use 200- to 300-grit sandpaper). For water-thin CAs, the surface should be smooth to about ten-thousandths of an inch.
- CA glues are more susceptible than are most adhesives to bond spoilage from dirt and oils. Many household adhesives, such as epoxy, are fairly tolerant of small amounts of contaminates, but CAs prefer it spotless. You can clean surfaces to be bonded with isopropyl alcohol and, when necessary, a solvent like acetone (use sparingly on plastics, as acetone melts most plastic). You may also add a tiny drop of acetone to the opposite surface to be joined to increase curing time.
- The best CA bond uses a minimum of material, applied to only one surface, not both. Use too much and the bond is considerably weakened. The squeeze bottles and tubes used for consumer CA products don't allow for accurate metering of the liquid, so more often than not, too much glop gets applied, even when you're being careful. Professional CA glue applicators are available, but can be expensive.
- Though many CA manufacturers claim otherwise, experience has shown that a CA glue bond can dry out and become brittle after merely a year or two. A joint that used to hold several hundred pounds may break with only a few pounds of exertion. This is especially true of bonds made without special metering applicators.
- Unless you purchase a special industrial formulation, avoid exposing any CA glue joint to heat or sunlight. The bond will become very weak and may even spontaneously come apart.
- CA glue, in the bottle or tube, has a relatively short shelf life. The optimum shelf life depends on the mixture and storage temperature, but don't expect longer than 12 to 18 months. If you have CA glues older than that, you might as well toss them, because their ability to provide a strong bond is

likely diminished. Keep the unused portion in a cool, dry place. Some hobbyists store their CA glues in the refrigerator, though I'm unconvinced this is a good idea. For one thing, the air in the fridge can be fairly moist, and moisture is an activator for CA glue.

- Finally, though CA glues set (or fixate) in less a minute, they are not fully cured for another one to four *days*. Although the curing time of many household glues can be accelerated by using heat—say, from a 100-watt bulb or even a heat gun—heat should not be applied to accelerate CA glue curing. You may, however, experiment with exposing the joint to the warmish, moist air of a room humidifier, or experiment with using acetone (as mentioned) as an activator.

Hot-Melt Glue

Most folks have seen hot-melt glue. This adhesive comes in stick form and is heated by a special gun-shaped applicator or even with an electric pot or crucible. Hot-melt glue is like a sticky wax: After being heated, the glue forms a pliable liquid. The liquid is applied to the surfaces to be joined. As a side note, for years paraffin waxes were commonly used for gluing, but this practice has largely gone by the wayside with the introduction of affordable plastic-based, hot-melt glue and guns.

Depending on the type of hot glue being used, melting temperatures range from 250 to 400° Fahrenheit. The cooling process doesn't take long—under a minute—and the glued joint is already strong. Additional strength comes as the glue cures while the pieces are securely held together. However, for most tasks, clamping is not necessary.

Though it may come as a surprise, hot-melt glue isn't glue at all, but plastic—or more specifically, a polymeric thermoplastic with a reasonably low melting point. Adhesion occurs as the molecules of the polymer contract and harden. A common polymer used for consumer hot-melt glue sticks is EVA. Many glue stick manufacturers use additives to enhance the properties of the EVA material.

Another hot-melt material gaining in popularity is PUR, a thermoset adhesive often used for industrial and automotive applications. PUR glues exhibit enhanced bond strength, and they are not as susceptible to remelting when exposed to heat.

The main benefit of hot-melt glue is that it sets quickly, yet yields a very strong bond. For most glues, the longer the setting time, the stronger the bond, which is why a 30-minute epoxy is better, all things considered, than a 5-minute epoxy. But the problem with slow-setting adhesives is that you either need to clamp the pieces together while the joint hardens, or you need to sit there holding everything in place. For some jobs, neither is practical.

Conversely, hot-melt glue sets in a few seconds to a few minutes, depending on the glue stick, the ambient temperate, the temperature of the surfaces being bonded, and several other factors. Full curing takes about 24 hours, but once set, the joint can be handled and, as mentioned, clamping is not usually necessary.

Using the Hot-Glue Gun

For most tasks, hot-melt glue is applied using a glue gun and glue sticks; both are available at hobby stores, arts and crafts stores, and most home improvement stores. The cost is under $20 for a gun and a half-dozen sticks. You can get a regular-size glue gun for bigger household jobs (see Figure 4-6), but a smaller craft gun is also available. You may wish to purchase both, as the larger gun is not well suited to applying small amounts of glue.

The glue is heated to a viscous state in a glue gun and then is spread over the area to be bonded. Not all glue sticks have the same melting point and hardness. The typical glue stick from the hardware or craft store is a low-temperature glue stick and sets to a semirigid state (the glue is not rock hard when set). This type is acceptable for use with most plastics to avoid any "sagging" or softening of the plastic due to the heat of the glue. However, for higher-bond yields for metals, closed-grain wood, and heavier plastics, select a high-temperature glue stick.

The procedure for applying glue to surfaces to be bonded is straightforward:

1. Prepare the surfaces to be bonded. Surfaces must be clean and dry.
2. If the surfaces are smooth, rough up one or both with 100- to 150-grit sandpaper.
3. If a glue stick is not already loaded into the gun, do so now.

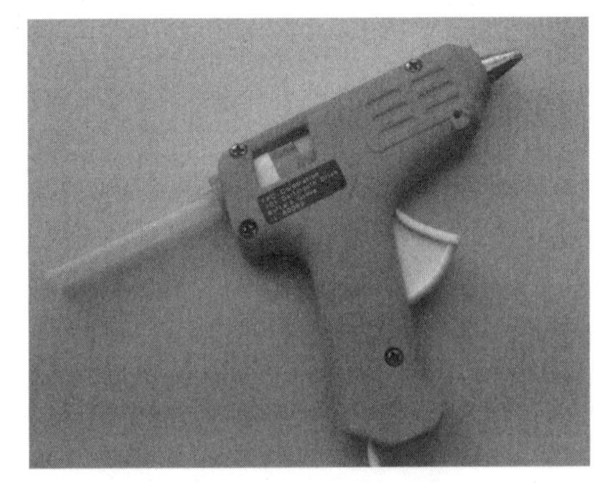

Figure 4-6 Hot-melt glue guns work by heating a polymer plastic to its melting point. A bond is created when the plastic cools, and its molecules contract.

4. Plug the glue gun into the electrical outlet and allow it to come to a proper operating temperature.
5. Clean the tip with a paper towel, as needed. Careful! The tip and any liquefied glue can be quite hot.
6. Test the glue application on a paper towel. The liquefied glue should come out readily. If flow is insufficient, look for an obstruction, or allow the gun to heat up some more.
7. When the gun is ready, apply a bead of hot-melt glue to one surface to be joined, as shown in Figure 4-7. Do not overapply—with hot-melt glue, less is more. If you need to apply glue to a large surface area, use a zigzag or spiral pattern to spread out the glue (see Figure 4-8). Avoid applying the bead closer than about ½" from the edge of the surface to be joined.
8. As quickly as possible (10 to 15 seconds, no more), bring the opposite surface into contact and apply pressure to spread the glue. If possible, while depressing, gently rotate the joint 5 to 10 degrees, and then realign as needed. This helps to spread the glue.
9. If any excess glue oozes out from the joint, wipe it up promptly with a paper towel. Don't try to remove it with your bare fingers . . . the glue is still very hot!

The better glue guns come with additional applicator tips or have them available for separate purchase. The standard nozzle applicator tip is acceptable for applying a bead of hot-melt glue, but other applicator tips are better suited to specific tasks. For example, applicator tips are available that apply an even, thin coat of hot melt ½" to 2" wide. These tips are ideal for gluing laminates, applying thin metal over wood, and similar tasks.

Figure 4-7 Apply a bead of hot-melt glue to one of the surfaces to be joined . . . as with most adhesives, a little goes a long way.

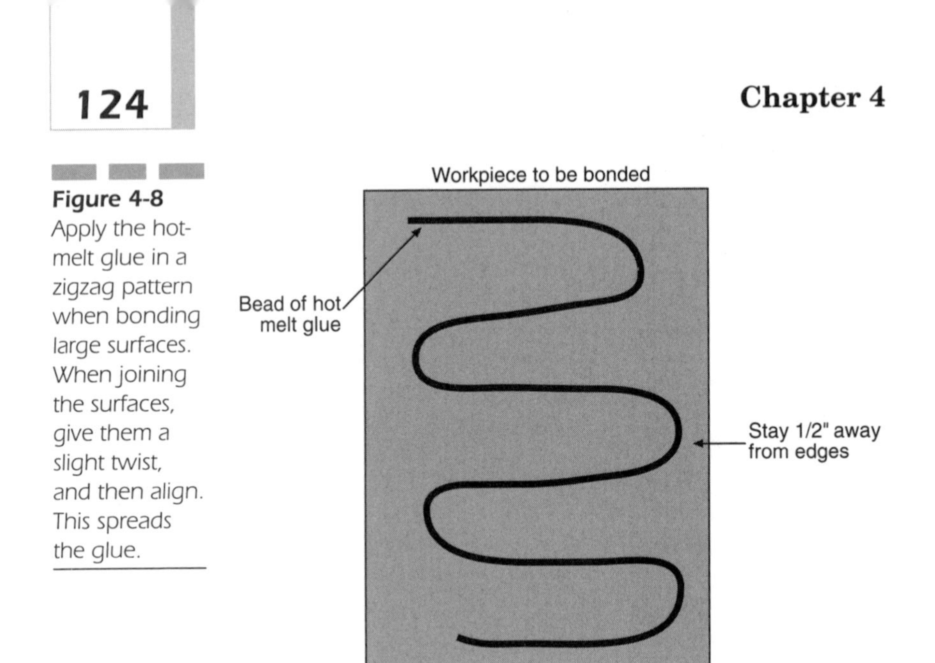

Figure 4-8 Apply the hot-melt glue in a zigzag pattern when bonding large surfaces. When joining the surfaces, give them a slight twist, and then align. This spreads the glue.

Using Hot Melt as a Casting Material

Hot-melt sticks can be used as casting material for making your own small plastic parts, such as gears, small tires, or bumpers. The glue sticks or glue pellets are readily melted, and the hot glue can then be poured into a mold. A heat pot or crucible is ideal for this. You can buy these from any school laboratory supply company. Or you can melt the glue sticks or pellets in a double-boiler pot, or, in a pinch, a regular cooking pot over an electric—but not gas—stove burner.

Select the glue-stick material that best suits the part you are casting. For those parts that must be flexible, select a softer glue stick (bend the stick to test it). Other parts may require a harder material.

Here are the general steps:

1. Create a mold using plaster, resin, or metal. Avoid latex or similar flexible molds because they can be affected by the heat of the hot melt. For fully enclosed 3D shapes, you'll need to create a two-part mold. Because the glue will set up fairly hard, avoid undercuts in the mold.
2. Apply a suitable release compound, such as a nonorganic food spray. Don't apply too much release compound, or the hot-melt plastic may be adversely affected. Don't use a silicone spray.
3. Melt the glue sticks in the heat pot. Typical softening point for low-temperature hot-melt glue is 180 to 265° Fahrenheit; it's 385 to 425° Fahrenheit for high-temperate hot melt. The viscosity of the material is reduced at higher temperatures. A molecular breakdown occurs as

temperatures exceed the flashpoint of the material (it varies depending on the glue stick). This will ruin the adhesive, so avoid too much heat.

4. Pour the melted hot glue into the mold.
5. Allow several minutes for the glue to cool and set before removing the piece from the mold.

Obviously, because the parts are made with hot-melt glue, they are susceptible to heat (unless you use PUR sticks). You can use this to your advantage if you need to join parts; they can be welded by carefully applying heat at the joints.

3M Hot Glue Comparisons

The 3M Company is a major manufacturer and supplier of hot-melt glues for both industry and consumers. Hobby and craft stores may sell all-purpose 3M glue sticks, but the better hardware stores and industrial supply outlets open to the public offer a wide assortment from which you can choose. Tables 4-1 and 4-2 list several 3M hot-melt glue stick products useful in constructing amateur robots.

Notes about both tables:

- The softening point is the temperature at which the material softens but is not yet in a liquid state.

Table 4-1 Low temperature (applied at 276° Fahrenheit)

Stock Number	Typical Applications	Color	Softening Point (°F)	Elongation %
3755	Paper, cardboard, corrugated chipboard; applied as thin bead	Clear	157	400
3762	Cardboard, wood	Amber	205	300
3776	Plastics, wood, light-gauge metal	Tan	184	600
3792	Heat-sensitive plastics (e.g., expanded rigid PVC), coated paper, wood	Clear	178	125

Table 4-2 High temperature (applied at 350 to 385° Fahrenheit)

Stock Number	Typical Applications	Color	Softening Point (°F)	Elongation %
3738	Cardboard, softwoods, light metals	Tan	220	1300
3762	Cardboard, paper	Tan	201	400
3764	Plastics, including acrylic, polycarbonate, and polypropylene; flexible at low temperatures	Clear	190	625
3789	Plastics, vinyl, and wood; high performance bond for plastics	Brown	270	600
3796	Light metals	Tan	240	930

- Elongation % indicates the relative flexibility of the set adhesive; the lower the percentage, the harder the material. You can easily judge the flexibility of the glue by bending the unused stick. The deeper you can dig in your fingernail, the softer the more flexible the glue.
- Typical colors are as indicated, but some materials are available in other colors.
- Most 3M hot-melt products are available in standard 2" (TC) and 8" (Q, or *Quadrack*) lengths.

Considerations for Hot-Melt Glue

Hot glue is an ideal adhesive for many jobs, but like everything, it has its drawbacks. Consider the following when using hot-melt glue:

- When using traditional glue sticks, the adhesive can be resoftened if exposed to heat. Do not use for joints near motors, sources of friction, or heat-dissipating electronics components. Otherwise, you could end up with a gooey mess.

- Roughen up plastic surfaces before trying to bond them. Plastics with a smooth surface will not adhere well when using hot-melt glue, and the joint will be weak. More than likely it will break off with only minor pressure.
- Hot-melt glue is viscous, and doesn't "thin out" when it sets, the way a water-based glue does. Remember to compensate for the extra bulk needed by the glue and provide space as required. (See Figure 4-9.)
- Most consumer-grade hot-melt sticks are susceptible to degradation from ultraviolet light. Avoid using them for outdoor robots.
- Though traditional hot melt is a thermoplastic, and therefore it can be heated and cooled to change its state, repeated heating and cooling will eventually degrade the material. Additionally, the molten plastic more readily attracts dust and dirt. It's best to discard once-heated hot melt, in case it has become contaminated.
- As with most any adhesive, the joints to be bonded should be free of dirt and oil, and they should be bone dry. Otherwise, a poor bond may result.
- Finally, though the hot-melt glue is not heated to extreme temperatures, it's still rather uncomfortable to have a drop of melting ooze land on your arm or leg. Exercise care when using the glue gun and keep it away from children. The low-temperature gun and glue sticks are safer (though the glue resets more quickly) but can still burn skin.

Some Favorite Adhesives

Everyone has a favorite adhesive, and everyone is fond of recommending his or her favorite to those who ask. There's even a line in the movie *Cannery Row* where Doc, the main character (played by Nick Nolte) recommends Duco, a household cement that's been popular for many decades.

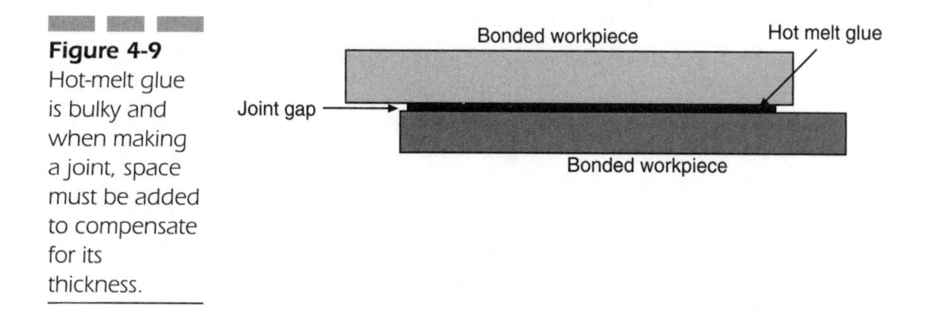

Figure 4-9 Hot-melt glue is bulky and when making a joint, space must be added to compensate for its thickness.

Like Steinbeck's Doc, I also think highly of Duco. It's a cellulose nitrate cement, with an acetone solvent, that dries clear, cures in under a day, and bonds china, wood, paper, fabric, and lots of other types of materials. I've used it for years for such things as gluing plastic servo hubs to foam wheels.

So there's one favorite. Here are some more; I polled some robot building friends and got some of their favorites as well.

- **Shoe Goo** Intended to repair worn or broken shoes, Shoe Goo is an all-purpose sealer and adhesive that sets up to a semispongy and quite resilient mass. Curing time is rather slow—one to five days—but curing can be accelerated by applying heat from a hair drier. For large or thick areas, apply several coats of Shoe Goo to build up the material, rather than one big clump at once.
- **Household GOOP** This product belongs to the large family of silicone or silicone-like adhesives that also include Plumber's GOOP, Crafter's GOOP, Sportsman's GOOP, and others. Household GOOP remains flexible even after curing. Most can also be peeled off with little effort, allowing for temporary bonds. Along similar lines is DAP adhesive caulk, another favorite of mine.
- **Pro Bond** This single-part polyurethane adhesive can be used to bond wood, plastics (most, anyway), metal, leather, paper, and other porous and nonporous materials. This is a good glue when you need to bond anything to anything.
- **Super Glue** Everyone calls CA glue *super glue*, but this is also a brand for a number of adhesive products. In this case, I'm talking about the product The Original Super Glue, available in a number of consistencies. Use the thicker gel formulations when the parts are not closely fitting. Super Glue Corporation sells products that come in tubes, bottles, and even pens. The parent company (Pacer Technology) also sells ZAP (popular among RC hobbyists), Bondini, and other brands of CA glues.
- **J-B Weld** J-B Weld is a company as well as a product. Most adhesive connoisseurs know J-B Weld as a bond-to-anything, two-part epoxy resin that contains fine particles of iron and steel. The adhesive bonds metal to metal, metal to plastic, and other variations of both porous and nonporous materials. J-B Weld is popular among auto restorers and weekend mechanics, and indeed, the product is available at many auto supply stores around the globe.
- **Barge All-Purpose Cement** Trust a shoe repair shop to know what contact cements are strong enough to tack heels onto boots. Barge products are not as easy to find as some others, but it'll glue most materials, particularly porous ones, with a most permanent bond. This is recommended when strength is critical. As with most contact cements, to use it you apply a coat to both parts to be bonded. Wait until the adhesive starts to dry (it'll get tacky to the touch), and then press the parts together.

- **Permatex** Representing a brand and a popular product, this stuff is really a gasket-making compound, but it can also be used as a sealant and flexible glue. Most adhesives cure to a hard mass, but Permatex is designed to stay flexible, such as silicone RTV. This allows you to more easily disassemble bonded parts should you ever need to. An advantage of Permatex is that it is very resistant to heat. It's commonly used in automotive and motorcycle repair. Other good products in the Permatex line include Permatex Cold Weld Bonding Compound. This two-part epoxy contains iron oxide filler and, once cured, can be sanded, drilled, and cut.

Alternative Adhesive Dispensing

Typical adhesives are dispensed from a tube or bottle. There are other methods, too. Earlier in the chapter you read about hot-melt glue guns, which are good examples of an alternative dispensing technology. There are also the following.

Glue Dots

Glue Dots is a trade name and is representative of a method of applying premetered adhesive. The dots are provided on a long roll and can be applied by hand or by machine. The dots come in various tacks: High tack provides permanent bonding, and low tack a provides temporary sticking place. The dots cure upon pressure.

Glue Stick

Glue Stick, yet another trade name, is representative of polymer-based products that smear on a jelly-like polymer adhesive from a self-contained applicator. Most consumer products are intended to be used to join paper to paper, but industrial stick adhesives are available for bonding metals, plastics, rubber, and other materials.

Adhesive Transfer Tapes

Adhesive transfer tapes are like ordinary adhesive tape, except only the adhesive portion—and not the tape backing—is left on the parts to the joined. 3M is a major manufacturer of adhesive transfer tapes. Their product and others like it are best applied using a special applicator that separates the protective backing from the tape, while laying the adhesive over the material to be joined.

Aerosol Adhesives

Aerosol adhesives are sprayed from a can, airbrush, or compressed air canister. The major advantage of aerosol adhesives is that they can be applied quickly in thin coats to a large area.

Clamping and Taping

Many adhesives do not bond instantly. Their setting time takes minutes, and in other cases, hours. During the setting time, it is critical that the bonded joint not be disturbed or else poor adhesion will result.

For very quick bonds—on the order of seconds—it is acceptable to manually hold the pieces together until they are set. Longer setting times may require clamping or taping in order to ensure

- Adequate pressure to seal the bond. The pressure of the clamp promotes full integration of the adhesive into the material. This mostly applies to porous materials but also affects some nonporous ones (such as plastics and metals) as well.
- No movement until the joint is set. If movement occurs, the adhesion may be greatly weakened.

Ordinary woodworking clamps, such as that shown in Figure 4-10, are adequate for most gluing tasks. If the clamp has an adjustable tension control, select a tension that exerts pressure but does not break or deform the material.

Taping the joint may also prove effective. This method is often used when using epoxies or acrylic adhesives with plastic. After applying the adhesive and mating the joint, tape is applied to keep the joint together. Masking tape works well in most situations, but if you need something stronger, white first-aid bandage tape—available in 1/2" and wider widths—can also be used. Be sure to use a tape that can be peeled off. You will find that tape removal is easiest after the adhesive has set but before it has cured. You may safely remove the tape after setting, if doing so will not cause undo tension on the joint. A sample taped joint is shown in Figure 4-11.

Figure 4-10 Woodworking clamps apply even and steady pressure to the joint until the adhesive has fully set.

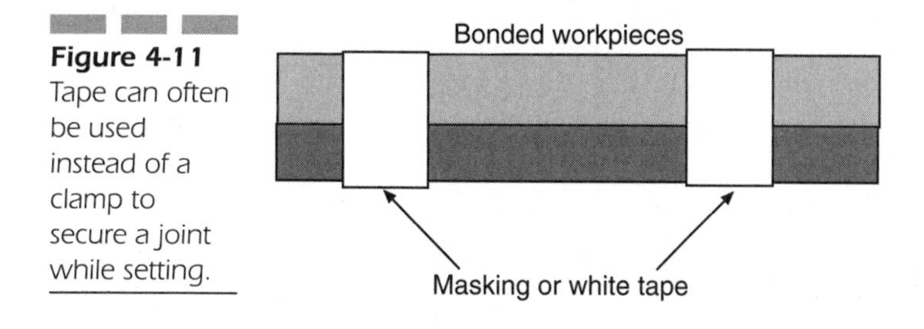

Figure 4-11 Tape can often be used instead of a clamp to secure a joint while setting.

Using Joint Reinforcements

Critical to the strength of any bond is the way the pieces are aligned and positioned. The weakest are butt joints (no jokes please), where two materials are bonded end to end. The reason: There is little surface area for the joint. As a rule, the larger the surface area, the more material the adhesive can join to, and therefore, the stronger the bond.

For a stronger joint, you will want to apply any of a variety of reinforcements that increase the surface area of the bond. Joint types and reinforcement techniques are explained in Figures 4-12 through 4-16.

Similar techniques can be applied to pieces that are joined at an angle. For example, Figure 4-17 shows a corner angle used to reinforce a 90-degree joint.

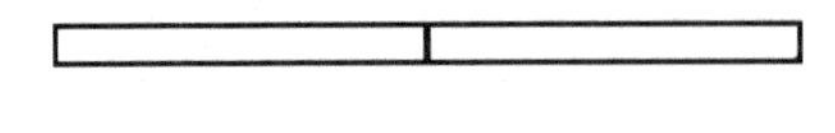

Figure 4-12
The typical butt joint provides minimal surface area for the strong bond. Avoid butt joints whenever possible.

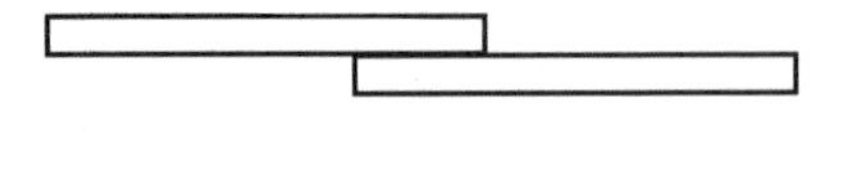

Figure 4–13
The pieces themselves can be overlapped instead of butting them end to end. This is not always feasible but is a quick and easy alternative when the option is available. You can readily adjust the amount of overlap as needed.

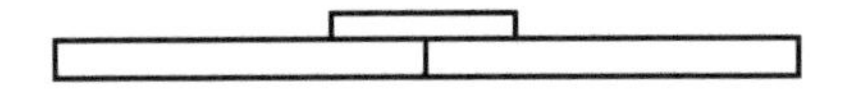

Figure 4-14
You can also overlap an extra piece of material along the seam of the joint. Use the widest overlap piece you can in order to increase surface area, apply adhesive to this extra piece, and clamp or tape until set as needed. You can reinforce it with small fasteners as needed.

Figure 4-15
The surface area of the joint can be increased by mitering the ends. This is most practical with materials that are 1/4" or thicker. The technique is particularly helpful when joining wood.

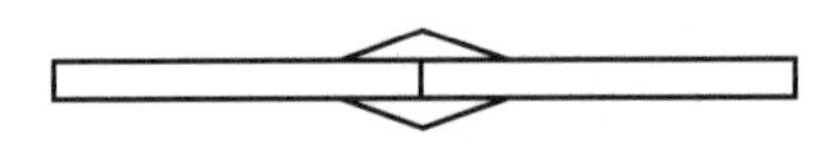

Figure 4-16
Gusset pieces can be used on the top and bottom. You can reinforce with small fasteners, as needed.

Figure 4-17
A gusset can be used to add strength to a 90-degree joint.

Joints with Molded Profiles

This chapter wouldn't be complete without mentioning a nonadhesive means of joining two pieces of material together, either end to end, or at an angle. This is accomplished using molded profiles. These are specially made plastic, rubber, or metal structures designed to couple two (or more) pieces of material.

In use, the material—typically a panel, bar, or tube—is inserted into the profile (see Figure 4-18). Some profiles provide a snug fit and nothing else is required to secure everything together. However, adhesives often can be applied to make the joint stronger and more permanent.

Molded profiles can be found at many home improvement stores, but selection may be limited. Online and local retail outlets that cater to the cabinetry industry are the best source of molded profiles, where you can select the color, material type, and other parameters to suit your needs.

In Summary: Selecting an Adhesive

With so many types of adhesives to choose from, it can be hard to select the right one. Table 4-3 summarizes the most common adhesive families, along with their pros and cons, and the bonds they are best used for. Table 4-4 provides various bonding recommendations for each major adhesive family.

Figure 4-18 Rubber or plastic profiles can be used instead of adhesives to join pieces.

Table 4-3 *Selecting an adhesive*

Adhesive Type	Pros	Cons	Best Used For
Acrylic	Strong bond, flexible	Primer and mixer required	Bonding dissimilar materials, such as metal to plastic, or foam to wood
Contact cement	Very fast adhesion	Careful assembly of parts required; toxic fumes	Laminating flat substrates
Cynaoacrylate	Good adhesion to rubber or plastics	Poor gap-filling, poor impact resistance, hard to accurately dispense with consumer tubes and bottles	Bonding porous and nonporous materials that are not subject to impact
Epoxy	Strong bond when prepared and applied correctly	Toxic fumes, curing sensitive to temperature, must be mixed correctly	Bonding all materials except silicone, Teflon, and other slippery materials
Hot melt	Readily available, fast setting, no harmful fumes	Can weaken under heat, poor impact resistance, accurate metering of adhesive difficult with consumer guns	Bonding wood, plastics, and light metals; use a high-temperature glue stick for a better bond with heavy materials
PVAC	Commonly available household adhesive	Not for use when both materials to join are nonporous	Bonding porous to porous and nonporous bonds (wood to metal, foam to plastic)
Solvent cement	Extremely strong bond in plastics and rubber	Requires matching the solvent to the material	Bonding plastic and rubber
Silicone	Remains flexible after curing	Toxic fumes, low strength	Bonding rubber and plastics; creating semiflexible seals

Table 4-4 *Recommended adhesives (by bonding material)*

	Bonding to			
Adhesives	**Metal**	**Plastic or Foam**	**Rubber**	**Wood**
Acrylic	Recommended	Acceptable	Not recommended	Acceptable
Contact cement	Acceptable	Recommended	Not recommended	Recommended
Cynaoacrylate	Acceptable	Recommended	Acceptable	Acceptable
Epoxy	Recommended	Recommended	Acceptable	Acceptable
Hot melt	Acceptable	Recommended	Not recommended	Acceptable
PVAC	Not recommended	Acceptable	Not recommended	Recommended
Solvent cement	Not recommended	Recommended	Acceptable	Not recommended
Silicone	Acceptable	Not recommended	Acceptable	Not recommended

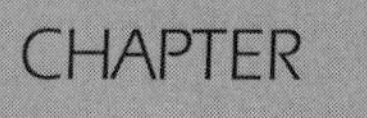

5

Mechanical Construction Techniques

The choice of materials is but one part of successful robot building. Another is how the materials are connected together. When you think about it, the same is true of any structure: A building can be reduced to a mound of concrete, steel, and glass if its construction is not sound.

In this chapter, we discuss numerous approaches to mechanical construction, from simple snap-together plastic parts, to hardcore fabrication using hand and power tools. You'll find additional helpful construction tips in Chapter 4, "Selecting and Using Adhesives," and in Chapter 6, "Robot Construction Hardware."

Temporary Versus Permanent Construction

Certain robot projects do not require permanent construction. Some robots are designed for limited testing and use, and therefore their construction doesn't need permanent fastening. Temporary constructions can often be built much faster than permanent constructions.

"Permanence" is actually "relative permanence" and is based on the size, weight, and use of the robot. A robot that is not made to carry heavy weight or bash into other robots can use lighter construction techniques, simply because there won't be as many external forces trying to tear it apart. Conversely, a robot designed for a rugged, outdoor environment, or a death-match competition may require more than simple nuts-and-bolts construction. For instance, these may need steel-reinforced welding to prevent parts from coming loose.

An obvious conclusion is that you must consider the robot and its environment when determining the best construction technique to use. That part is easy; the hard part is knowing which technique is advisable. When will glued joints give out? How much stress can that 6-32 machine screw withstand before snapping?

Alas, these are engineering problems that would fill a book of its own, and they require an in-depth knowledge of materials and production techniques. Such discussion is well beyond the scope of this book, but if you're interested, a list of engineering resources can be found in Appendix A, "Resources."

Nevertheless, for amateur robotics there is little need to overengineer its design by analyzing every joint and cross-member. Rather than getting bogged down by esoterica, we can concentrate on the more fun aspects of robot building and apply some simplistic rules for categorizing construction techniques with the types of robots they best represent. See Table 5-1 for a comparison of common construction methods.

Note that many robots are built using more than one construction technique. Bear in mind that the robot is only as strong as its weakest part. Additional information on these construction techniques is shown in Table 5-1.

Table 5-1 Common construction techniques

Construction Technique	Applicable To	Not Applicable To
Snap-together parts	Small toy-based robots (LEGO or K'NEX)	Designs that require permanence, unless parts are glued
Adhesive tape (electrical tape, duct tape, and so on), double-sided foam tape, standard-strength hook and loop (such as Velcro)	Test designs; temporary constructions; lightweight materials and components; heavy-duty hook and loop (industrial Velcro, 3M Dual Lock) applicable for larger weights and stresses	Long-term constructions; material or component weights of over a few ounces
Glues	Dependent on adhesive, can be used on robots of up to several pounds	Heavier metal robots; components that may receive heavy, sudden shock
Machine screw fasteners	General-purpose robots, but size and weight limitations apply	Very high-strength and high-impact applications (heavy combat robots); outdoor use unless stainless steel fasteners are used; larger and heavier robots require large bolts, which can add to weight (welding is advisable as an alternative)
Rivets	Same as machine screw fasteners, but rivets provide weight savings	Constructions that may need disassembly; high strength applications
Soldering and brazing	Small to medium-size and small to medium-weight robots	Very large or heavy robots where considerable impact or shock may occur
Welding	Large and heavy robots, particularly those used in rugged environments	Small or low-cost robots

Construction with Snap-Together Components

Plastic construction toys like LEGO (see Figure 5-1), K'NEX, and MEGA BLOKS offer a quick and easy way to build small tabletop robots. Unless you're going after an unusual design, no cutting or drilling is involved—simply pick the piece you want to use and snap it into place. Chapter 1, "The Basics of Robot Bodies," discussed a number of construction toy sets that are applicable to robot building, so it won't be repeated here.

Although it is possible to build a complete robot with only a certain type of snap-together construction toy, you are in no way limited to doing so. In fact, you'll find your robot creations can be better and even stronger if only certain components are constructed from snap-together parts. You can combine some LEGO pieces with a small block of wood or plastic, for example, cementing everything with glue, tape, or even nylon tie-wraps. Or you may use the flexible plastic beams from a K'NEX set as linkages in a walking robot. You may need to cut and drill these parts to better fit your robot designs, but in most cases, the work isn't hard because the parts are made of plastic and can be worked with ordinary shop tools.

Snap-together components are by their nature temporary. They are made to be taken apart and reused. This may be your aim with your latest robot creation. Also bear in mind that temporary constructions can come apart when you don't want them to, especially if the robot is mishandled, takes a fall from the workbench, or bangs into objects or other robots.

If you need a more permanent construction, apply a suitable glue to the snap-together parts. The best glue to use depends on the parts and the type of joint you wish to make (see Table 5-2).

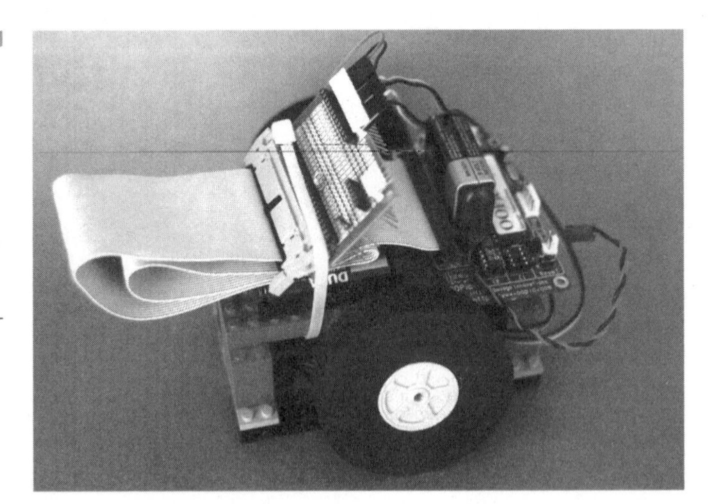

Figure 5-1 LEGO blocks and other pieces provide a quick and convenient means to build simple robot platforms.

Table 5-2 Recommended glues for plastic construction toys

Construction Parts	Temporary	Permanent
LEGO and K'NEX	White glue	ABS plastic solvent cement; two-part epoxy
MEGA BLOKS, plastic models (model cars or airplanes)	White glue, or low-temperature hot-melt glue	Plastic model-building (polystyrene) solvent cement; two-part epoxy
Robotix, Fischertechnik, Construx, and most other plastic construction toys	Low-temperature hot-melt glue	ABS-PVC solvent cement; two-part epoxy

- For a strong but less permanent bond, use only small amounts of solvent cement or epoxy.
- High-temperature hot-melt glue provides a good middle ground between temporary and permanent constructions. Use sparingly if you wish to disassemble the parts later. The glue can usually be peeled off.
- Flexible adhesives, such as Shoe Goo, Household GOOP, or any of a number of silicone-based RTV adhesives, also make for strong yet temporary bonds.

Though snap-together parts are most often used in robotic constructions with or without adhesives, it is also perfectly acceptable to use other binding techniques with them, including machine screws and other fasteners, double-sided foam tape, or nylon tie-wraps. By no means are you limited in any way to a single method to lash the parts together. As the variations are endless, we'll just leave the discussion at that and let your creativity come up with interesting alternatives.

Construction with Sticky Tapes, Ties, and Hook and Loop

Tape, ties, and hook and loop are used to produce temporary and semipermanent constructions. The level of permanence depends on the product, the surface area (larger tape means a higher bond), and the material of your robot.

Unless otherwise specified, the following discussion applies to consumer-grade sticky tapes and hook-and-loop material. Industrial-grade products offer greater holding power but are more difficult to find. Often, they are available only in bulk from specialty industrial suppliers, such as Grainger's and McMaster-Carr.

Sticky Tapes

Sticky tape is a broad family of products that have an adhesive on one or both sides of the tape. Sticky tape is cheap, is easy to use, and bonds nearly instantly to the surfaces to which it is applied. Although sticky tape makes for handy construction material, remember that the tape adhesive is gummy and can leave residue on the parts. Denatured alcohol can be used to remove the residue as required. (But test first to ensure that the alcohol doesn't dissolve the parts of your robot you want to keep!)

Although tape adhesive may be strong, few products are meant to be permanent. This may be exactly what you're after, but if you're not, tape that unravels or causes parts to fall off can be a source of frustration.

More important is that the adhesive used with most sticky tapes is not dimensionally stable. Under stress and loads—such as a drive motor—parts may shift under the adhesive. This so-called "creep" can cause components of the robot to become misaligned over time. For this reason, I recommend using sticky tape for noncritical applications only, such as mounting battery holders or fastening plastic panels over a metal frame.

Double-Sided Foam Tape A common staple in any robot builder's workshop is a roll of *double-sided foam tape*. This tape is composed of a layer of springy foam, usually either 1/32" or 1/6" thick, and from 1/4" to over 1" wide. The tape is coated with an aggressive adhesive on both sides. To use, you peel off the protective paper and apply the tape between the parts to be joined. The adhesive is pressure sensitive and cures to a strong bond within 24 hours.

There are many classifications of double-sided foam tape, with ordinary consumer tapes only marginally useful in robotics. The adhesive used with these tapes is designed for lightweight use only. In fact, some double-sided foam tapes are engineered with an adhesive that never fully cures. It stays gummy so that the tape can be more readily removed from walls.

Industrial-grade, double-sided foam tape is available at better hardware stores, as well as at industrial supply mail-order outlets, such as Grainger's and McMaster-Carr. Although these tapes cost more and come in longer rolls, they are significantly stronger. Unless you plan on using a lot of double-sided foam tape, purchase the smallest roll you can, because the adhesive does dry out over time (from one to two years). Keep the roll in a sealable sandwich baggie for a longer shelf life.

Duct and Electrical Tape *Duct tape* is known for its highly sticky adhesive. It's made to seal air ducts for ventilation systems—hence its name. By design (to accommodate expansion and contraction), both the tape itself and the adhesive are flexible. This makes duct tape ill suited for mounting motors or other components that must be kept aligned. Though some duct tape is plastic, much of it is made of paper and is not waterproof. In fact, duct tape can hold water, which can lead to a total breakdown of the adhesive backing.

Less sticky is *electrical tape*, composed of a *polyvinyl chloride* (PVC) or similar, flexible plastic, coated on one side with a nonconductive adhesive. Designed to insulate wire connections, electrical tape can also be used to bind materials together. For best results, the tape should always wrap around itself, preferably with a stretch to keep it taught.

Masking and Box-Sealing Tape Tape for painting is often referred to as *masking tape*, because it's intended to mask off an area to protect it from the paint. Most masking tapes are paper with a nonpermanent adhesive. The paper may or may not be coated. Consumer-grade masking tapes are best suited for very lightweight applications, such as keeping a bundle of wires together. Industrial-grade masking tapes are tougher and have various grades of adhesives. Gaffer's tape, available from music stores and other outlets that cater to the local disc jockey trade, offers a more robust paper backing and adhesive. It's available in a rainbow of colors.

Box-sealing tape is used in cardboard packaging. The tape is plastic, and the adhesive varies from light duty (common with consumer tapes) to very permanent. An ideal use of the tape is applications where the adhesive backing can contact itself, as this provides a very strong bond. The primary disadvantage of box-sealing tapes is that most are not very flexible and will tear.

Gummy Transfer Tape Here's an old trick: Suppose you want some "stickum" but can't use tape. One way is to apply a coating of rubber cement over some waxed paper. Wait half a minute for the cement to congeal, and then gently roll up the residue into small clumps. You can now stick the stickum where you need it. One disadvantage of this technique is that rubber cement tends to dry out, and its stickum doesn't stay sticky for long.

Transfer tape is an option. This stuff, which is also called *unsupported adhesive tape*, looks like double-sided tape, but it's engineered to leave its sticky residue, and no tape. It's meant for such jobs as electronics production, where workers apply small dabs of sticky substances to hold down wires and components. Transfer tape, such as that made by tesa AG in Germany, is applied by first placing the material as you would any other tape. You then peel off the waxed paper, and what's behind is a layer of very sticky goo. Unlike rubber cement, transfer tape stickum remains tacky and semiflexible.

Transfer tape is available from art supply stores, adhesive specialists, and electronics production supply outfits. Here are some online sellers of adhesive transfer tape (note that some tapes work best when used with a specially made dispenser):

Curry's Art Store Limited ***www.currys.com***

Hillas Packaging ***www.hillas.com***

Jerry's Artarama ***www.jerryscatalog.com***

ULINE ***www.uline.com***

Self-Cling Stretch Film Some plastics are known for their cling. These include polyethylene and PVC. These materials provide extra interlock strength when slightly stretched. Depending on the plastic, the cling is the result of *stiction*, caused when the smooth surfaces of the material come into contact. Other plastics, such as Saran Wrap (technically a copolymer of vinylchloride and vinylidene chloride), are treated to enhance their ability to retain a static electrical charge. Similar products are Reynold's Plastic Wrap (a PVC) and Glad Wrap (polyethylene).

Saran Wrap and similar products are too thin to be really useful in robotics. But other plastics are available in varying thicknesses, and these can be used when you need a sticky tape that doesn't use an adhesive. One choice is *plasticized PVC*, available in roll or sheet form, in thicknesses from 150 microns (0.15 mm) to over 2 mm. This material is commonly used to make signs for windows, and so is readily available at sign supply outlets. A local sign maker may sell you scraps if you ask nicely. Be sure it is the type that will cling to itself, not just to glass.

To use the plastic, cut it into usable pieces and clean with denatured alcohol. Apply the tape by rolling it around itself. The cling will not be particularly strong, but should be adequate for such tasks as bundling wires and temporarily holding small parts together for subsequent gluing.

Shrink Film *Shrink film* is designed to shrink upon exposure to heat. It's primarily used for packaging goods for resale. The film is available in tube, sheet, tape form, and can be used with a hair dryer on high heat or a heat gun especially made for the job. Shrink film is useful in robotics as a means to protect components, to create unique homemade sensors, and to bind wire bundles.

Shrink film and tape can be found at any packaging outlet, though if you buy some, you'll have to purchase fairly large quantities. See if you can obtain a small sample or leftover pieces from a larger roll.

Blenderm: Sticky Stuff Without Gummy Residue 3M's *Blenderm* is a waterproof adhesive bandage that is intended to be used on people, but is also pretty handy with robots. The semiclear tape is available in different widths up to 2", and sticks to practically anything. Yet it doesn't leave much gummy junk behind, and it stays flexible for an eternity. Possible ideas: Use it to construct hinges or movable flaps.

You can sometimes find Blenderm at the local drug stores, but be wary of the cheaper brands. You want the real stuff, so make sure it says *3M* and *Blenderm*. The typical online sellers of Blenderm are medical supply outfits. Here are some to get you started:

Elite Medical ***www.elitemedical.com***

Medical Supply Company ***www.medsupplyco.com***

Global Drugs ***www.globaldrugs.com***

Hook and Loop

Velcro was discovered when its inventor noticed how burrs from weeds stuck to the fur of his dog. The construction of Velcro is a two-part fabric: One part is stiff (the burrs) and the other soft (the dog). Attach them together and they stick. The term Velcro is a combination of the French words *velour* and *crochet*.

Velcro is a trade name for a kind of hook-and-loop fastener, and the Velcro Company probably sells more of it than any other company does. It's available in a variety of sizes and types, from the ordinary household Velcro you already know about to heavy-duty industrial strips that can support over a hundred pounds. Figure 5-2 shows some Velcro in action.

Among the most useful hook-and-loop products is the continuous strip, where you can cut what you need to length. The strip comes in packages of 1 foot to several yards, in any of a number of widths—½" and 1' wide are the most common.

Figure 5-2 Velcro is often used to mount parts, such as small casters, to a robot base.

The strips come with a peel-off adhesive backing that allows you to directly apply the hook and the loop to the parts you wish to fasten. If the adhesive is not strong enough (which is sometimes the case), you can reinforce the material with a heavy-duty epoxy, screws, staples, and so on.

Although Velcro may be the best-known hook-and-loop material, it is, by far, not the only one available. A great alternative is 3M Dual Lock (see Figure 5-3), a unique all-plastic strip that is composed of tiny tendrils. The tips of the tendrils are ball shaped; when the material is brought into contact with itself, the balls intertwine, providing a very strong hold. Unlike Velcro, Dual Lock has no separate hook-and-loop component. It sticks to itself.

Dual Lock is not quite as well-known as Velcro, but a number of retail stores carry it. A few online specialty stores, such as Budget Robotics, sell it by the foot.

Plastic Ties

Intended to hold bundles of wire and other loose items, plastic ties can also be used to hold things to your robot. The tie is composed of a ratcheted strip and a locking mechanism. Loop the strip into the mechanism and pull the strip through. The locking mechanism works in one direction: You can tighten the strip, but can't loosen it (this applies to most plastic ties; some have a releasable lock).

Plastic ties are made of nylon and are very strong and durable. They're available in a variety of lengths, starting at 100 mm (a little under 4") to well over 12". Save the larger ties for the heavy-duty jobs. For most applications, the 4" to 6" lengths will work fine. You can anchor the tie into a hole drilled for the purpose or use one of several mounts specifically designed for use with plastic ties

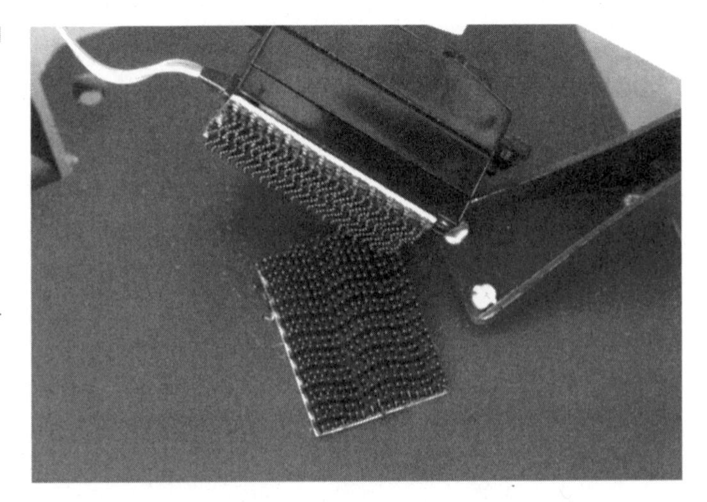

Figure 5-3 Dual Lock is similar in application to Velcro but uses a rough-surfaced plastic that can be joined to itself.

(see Figure 5-4). I prefer mounts designed for use with hardware fasteners. Drill a hole for a 4-40 or 6-32 machine screw, and secure the mount with a screw and nut.

Hardy Construction Techniques

The remaining construction techniques require a deeper level of discussion than can be afforded in this chapter alone. These include working with glue, soldering, brazing, welding, and assembling robots using nuts, screws, and other fasteners. Following are brief reviews of these techniques, along with references to other chapters that provide additional detail.

Construction with Glue

Glue is important stuff in any building endeavor. Glue is used to simplify construction and can reduce the weight of the finished piece compared to rivets or fasteners. When properly applied, glued joints can be equally as strong, and in some cases, stronger than the base material itself. Parts of airplanes are glued rather than riveted or fastened together, both as a cost-savings technique and to make the plane lighter. Chapter 4 is devoted to this subject, given the importance of glues in robotics.

Figure 5-4 Use a mount to anchor a plastic tie. Mounts are available with adhesive backing (they're the weakest) or with holes for securing with hardware fasteners.

Construction with Soldering, Brazing, and Welding

Soldering, brazing, and welding are techniques used to join metal (also plastics, using a suitable plastics welding tool). Each of the three is distinct, though they all use heat to soften or melt one or more metals. Of these techniques, welding is probably the most widely used. Heavy-duty robots are commonly welded together rather than assembled using fasteners. Soldering and brazing are more frequently used to produce individual components of a robot, such as a metal scoop for a sumo 'bot. See Chapter 10, "Going Further," for some additional details on soldering, brazing, and welding.

Construction with Mechanical Fasteners

Mechanical fasteners include nuts, screws, bolts, and other hardware used to hold pieces together. Fasteners are favored because they are cheap and easy to get, and unlike adhesives or even welding, they don't require that you exactly match the materials to be joined. Most fasteners can also be undone, so you can disassemble the robot if need be. This last trait is actually quite handy in academic environments where one or two robots must be shared between many students. After one group has built and experimented with the robot, it can be disassembled and is then ready for the next group. See Chapter 6 for a detailed look at fasteners suitable to the robotics art.

Producing Drill and Cut Layout Templates

Everything goes better if you have a plan. A layout template marks off the cuts you want to saw and the holes you want to drill. You can draw this layout directly onto the part you are crafting or produce the layout on paper and then transfer the layout to the material.

Layouts by Hand

The straightforward method of producing parts layouts is to hand draw them. The drawing can be directly on the part itself, or on paper, with the layout transferred using any of several techniques.

To produce the layout, use a sheet of ordinary, unlined white paper. Using a pencil, draw the layout onto the paper. A ruler or other drafting aid will help in making straight, accurate lines. If you wish, you can use graph paper ($^{1}/_{4}$" grid) to help with the layout.

In most instances, the paper template can be fixed to the material, as shown in Figure 5-5, whether it is wood, plastic, or metal. Use tape, a glue stick, or other temporary adhesive to hold the paper to the material. You can use the layout to punch pilot marks prior to drilling. A spring-loaded punch, available at any hardware store, is ideal for this job. After cutting and drilling is completed, peel the paper away from the material. For plastic and metal, any adhesive residue that is left can be cleaned off using denatured alcohol. For wood, the adhesive can be removed by a light sanding. Avoid the use of liquids as they may raise the grain of the wood.

The paper method is useful if you need to make several of the same part. Draw the layout once, and then have it copied on a plain-paper copier. To be sure the copier has reproduced the images at 100 percent, hold up the original and the copy to the light, noting any misalignment. Some copiers automatically apply a 2 percent (or so) reduction, and this can be compensated for on the better copiers by enlarging the image.

Occasionally, it is necessary to transfer a design from paper directly onto the material. If the material does not have a rough surface, you can transfer the design using a sheet of transfer paper, which is available at most arts and crafts stores. Transfer paper works just as well as old-fashioned carbon paper, but it

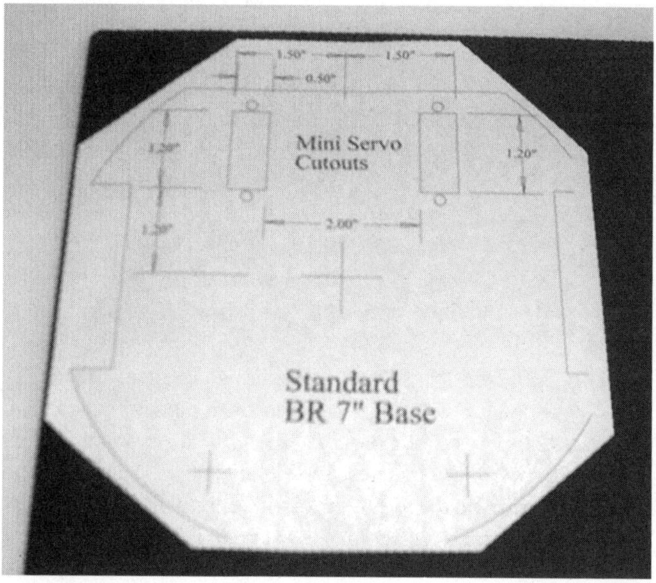

Figure 5-5 Paper templates make it easier to cut and drill material with accuracy.

isn't as messy. Unwanted transfer lines can be removed with a soft pencil eraser. Trace the design on the paper using a ballpoint pen. The tracing will appear through the transfer paper.

For materials with a rough or irregular surface, the pattern can be transferred using a scribe. A machinist's scribe is the appropriate tool for the job, but these can be very expensive. Most any sharp metal implement, such as a scratch awl or even a nail (with its tip sharpened), will work with plastics, aluminum, and other soft metals.

For harder metals, or for a more distinct transfer pattern, apply a thin coating of *layout ink* over the paper; then scribe the design into the ink. Layout ink is a common staple at any machinist supply outlet. Avoid the use of regular pen ink because it smudges easily and will make a mess. After cutting and drilling, the ink can be removed with a spirit solvent (such as paint thinner).

Layouts with Free and Low-Cost CAD Programs

One of the best ways to produce layouts for your robot projects is with a *computer-aided design* (CAD) program. CAD is more than merely a drawing program; it combines the ability to draw and edit geometric shapes with the layout precision needed for high-resolution drafting. The idea behind CAD is that not only can you draw a square, but you can draw a square that is precisely 1" × 2" or any other dimension that you choose. Absolute measurements are stored with the CAD file and, when used with the appropriate printer, produce highly accurate renditions of your drawings.

NOTE: *CAD programs are often referred to as 2D or 3D. A 2D CAD program can create a two-dimensional drawing. The layout on the drawing has height and width, but no depth. A 3D program can create a three-dimensional drawing that has height, width, and depth. Most 3D CAD programs can render 3D shapes using complex lighting and shading options. 3D CAD is not required for producing basic robot layouts.*

AutoCAD, from AutoDesk, is perhaps the best known CAD program and is used by thousands of engineers and designers worldwide. As with many commercial CAD programs, AutoCAD is frightfully expensive. If you're a student, you may qualify for a student discount, but even so, prices may be higher than what you'd like to spend for a tool that will see only occasional use in the robot lab.

An alternative is a free or low-cost CAD program. Several are available for download from the Internet. Although they may not compare with high-end commercial products like AutoCAD, they are more than sufficient for our application.

Among the leaders in low-cost CAD programs is IntelliCAD, a quasi-open source project by the IntelliCAD Technology Consortium. IntelliCAD is located on the Web at ***www.intellicad.org***.

(Note that *open source* doesn't strictly mean free. It means that programmers join forces to produce software that is made available to the public. Fees can be charged for open-source software. This is the case with versions of IntelliCAD from CADopia; this company offers and supports a commercial version of IntelliCAD.)

Viewed in brief, IntelliCAD is something of an AutoCAD clone, though this is an unfair description to both programs. Both IntelliCAD and AutoCAD offer features that are unique to each. Both use the same file format (DWG), and IntelliCAD can also run many AutoCAD add-in programs and *list processing language* (LISP) macros (some conversion may be necessary). Like AutoCAD, IntelliCAD supports both a menu-driven and command-line prompt.

I'll let the IntelliCAD Web site and documentation speak for themselves, and cut straight to using this tool to build robot layouts. Figure 5-6 shows IntelliCAD with a simple drill-and-cut template (note that the program supports hundreds of onscreen shortcut buttons not shown here, and note that for this example I'm using the CADopia branded version; the appearance of your version may be different).

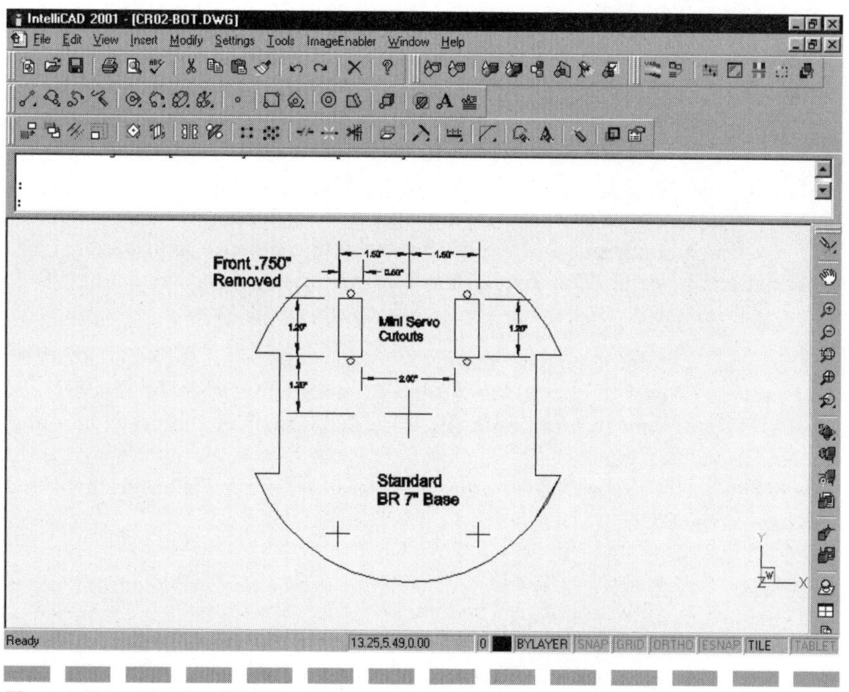

Figure 5-6 Use IntelliCAD to draw layouts for your robot designs.

Basic CAD Functionality Typical of CAD programs, IntelliCAD requires something of a steep learning curve to discover how to use its features. But basically, and for our purposes, the functionality of the program can be narrowed to the following:

- **Drawing setup** Here you define the drawing size and drawing scales (such as 1:1 or 1:12), the unit of measurement, grid size, and drawing resolution. For most robotic projects, you'll want a 1:1 scale, a grid of 1/4" or 1/8", and a resolution of two or three decimal places (down to the hundredths or the thousandths). Pick the unit of measurement (inches, millimeters, and so on) based on convenience and habit. For layout work, a resolution of two decimal places is satisfactory.
- **Drawing primitives tools** Only a few are used for typical layout drawings: line (or polyline), circle, and rectangle. Lines are used to mark cutting layouts. A polyline is a set of lines that share at least two or more vertices and is used whenever you want to cut out more complex shapes. Circles are typically used to denote holes for drilling. A rectangle or square is a closed polyline shape and can be produced using the line, polyline, or rectangle tool.
- **Editing and sizing tools** You can adjust the size and shape of the primitives by using the mouse or by entering values at the command-line prompt. The mouse is good for eyeballing the design, but the command-line entry of values is handy when you need accurate placement.
- **File saving and printing** Once done with the drawing, you can save it for future use or print it out. Any supported printer will do, such as a laser or inkjet printer. CAD programs don't have to be used with pen plotters anymore, as long as the drawing will fit onto the paper you've loaded into the printer. When using a laser or inkjet printer, make some sample shapes of known sizes, and then measure the resulting printout. You may need to make fine adjustments in the printing setup to compensate for minor variations in sizing. The manual for the CAD program you are using will explain how to do this, if the program supports this feature.

Drawings are placed on a *workplane*. With 2D CAD, a simple X and Y coordinate system is used to denote the origin of the drawing. With IntelliCAD, and most CAD programs, the origin is the lower-left corner of the drawing and is denoted as 0,0 (see Figure 5-7). The first digit is the X-axis; the second digit is the Y-axis. The values are in inches, millimeters, or other units of measurement that you have selected.

Benefits of CAD There are several benefits of using a CAD program to create cutting and drilling layouts:

- **Accuracy** With CAD, it's relatively easy to draw shapes with the *exact* size you want. No more guessing. You can precisely control both the size of lines, circles, and other primitives, and their spacing to one another. This can be done using the sizing and dimensioning tools, and with *snaps*, a feature supported by IntelliCAD and most CAD programs. Snaps force drawing elements to conform to known sizes and boundaries.

- **Drawing automation** If you need to produce a series of 20 holes around the circumference of a 6" circle, for example, tools provided by the CAD program make this easy. CAD programs support automation in different ways. Some let you write or record *macros* for repetitive tasks. Others offer features such as entity copying, which allows you to first select the drawing element you want to reproduce, and then indicate the number of copies and their distribution over the drawing area.
- **Editability** Designs can be readily and quickly altered, in case you need to make adjustments. This is sometimes necessary as you refine a design to make it better. Although some refinements can be made on the fly as you work with the finished robot, you may also wish to go back to the original design, make changes, and start over again. The edited version is also useful if sharing your designs with others.

A fourth benefit applies if you own, or have access to, a computer-controlled mill or router. A drawing you produce with the CAD program can be used directly to produce the layout on the mill or router. This requires the CAD program to support whatever file format is required of the mill or router tool. Most computer-controlled mill and router software can read a DXF file, which is a commonly supported file format of 2D CAD programs. We talk more about computer-controlled machining tools later in this chapter.

Other Free or Low-Cost CAD Options IntelliCAD is but one of several free or low-cost CAD programs. Most are available for the PC, though versions for

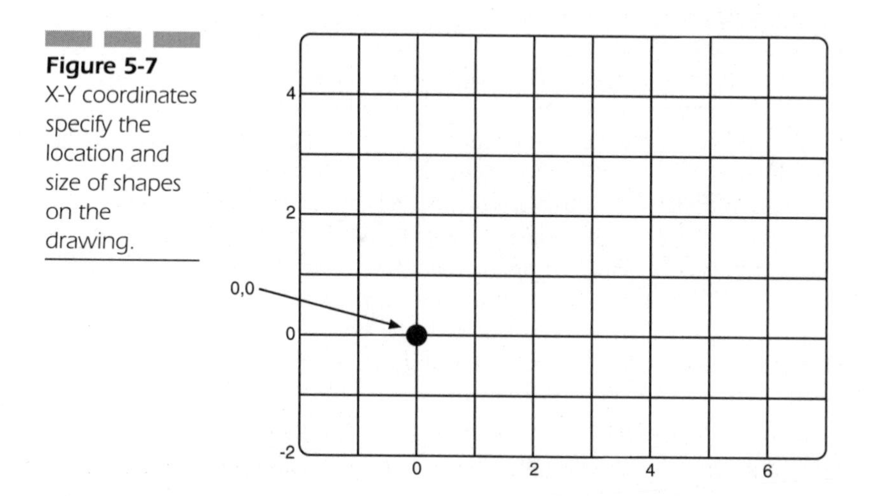

Figure 5-7 X-Y coordinates specify the location and size of shapes on the drawing.

Linux and the Macintosh are also available. Some are lite versions of higher-cost commercial applications, while others are shareware (pay if you use) and time- or feature-limited (but otherwise functional) demos. The following list is by no means exhaustive, and it should be noted that some of the programs—and even Web sites—may come and go. Always check Google.com and other search engines for current leads.

CadStd Lite (lite version) ***www.cadstd.com***

TurboCAD (low-cost standard and time-limited demo versions) ***www.turbocad.com***

DESI-III ***users.pandora.be/desi-iii***

Open Cascade (open source CAD and engineering tools) ***www.opencascade.com***

PowerCAD (open source CAD) ***sourceforge.net/projects/powercad***

vDraft (demo), ***www.softsource.com***

Additional CAD programs, utilities, and tools ***www.caddepot.com*** or ***www.cadalog.com***

Rough-Cut Builds and Quick Protos

Not every robot needs heavy-duty construction. Sometimes all that's required is a general idea that a particular design is workable. Rather than use the traditional construction techniques you'd apply to a permanent creation, you can build a rough-cut prototype of the robot and use it for testing purposes.

Cutting and drilling of the robot body is among the most time-consuming tasks. To reduce the time required to produce the prototype, select a material that is easy to cut and drill, perhaps even with hand tools. For small desktop robots that require only marginal strength in a temporary prototype, a material known as corrugated plastic (Gatorboard is a common brand name) is ideal for the job. The plastic is about 1/4" thick and can be cut with a knife or even heavy-duty scissors. A quick mock-up or prototype can be roughed out in just a few minutes using only simple tools. Some corrugated plastic is shown in Figure 5-8.

Corrugated plastic gets its rigidity from its fan-fold design. It's meant to be used as a backing for temporary outdoor signs, so it's not particularly hearty. If you need a stiffer substrate, you can use several layers of the plastic, sandwiching the layers at 90 degrees. This orientation increases the rigidity of the material.

Foam board (a.k.a. Fome-cor, a brand name) is likewise a good candidate for quick prototypes. This material, available at most craft and art supply stores, is constructed of foam, laminated on both sides with stiff paper. Foam board can be cut with a knife or small hobby saw, and holes can be drilled in it with a hand drill. Because the board is laminated with paper, you can use any of a number of paper glues to try out different designs. Use ordinary white glue if you don't mind waiting 30 to 60 minutes for it to dry. Otherwise, select quick-drying contact cement.

For designs requiring mounting small motors to the robot, consider the use of Velcro or similar hook-and-loop material. Fastening servos and other motors to lightweight substrates, such as corrugated plastic and foam board, may prove frustrating—the fasteners may pop through any holes you drill. Using hook-and-loop materials also allows you to more easily experiment with the placement of the motors and other parts. Most hook-and-loop material has a self-stick backing, but if the adhesive proves inadequate, you can reinforce it with staples (in the case of corrugated plastic or foam board) or a stronger glue.

Construction Process

Ideally, at one time or another in your life, you've used hand and power tools. Maybe you constructed a pipe rack for dad in shop class—never mind he doesn't smoke; it's the thought that counts. Or perhaps you fixed the deck in the back yard, or turned the basement into an entertainment den, complete with home theater.

Figure 5-8 Corrugated plastic is an ideal choice for quick, temporary prototypes.

In any case, although tool-working experience was taken for granted among the male population as recently as 20 years ago (females seldom were expected to have used tools), this presumption doesn't apply today. In fact, odds are, male or female, you may have graduated high school without ever taking a wood or metal shop class.

If you lack basic construction and tool use know-how, this book can't help you. Trying to teach the fundamentals of drilling a hole, or sawing a straight cut, would eat up pages we need to devote to the specifics of robot building. If you're lacking in shop skills, visit the library and check out any of the great beginner-level books on the subject. The Time-Life and *Reader's Digest* do-it-yourself books are particularly well done, with clear and concise illustrations.

Don't limit yourself to only reading. Try a small project, and practice. You might be surprised how much skill goes into drilling a good hole. After some practice, the process will become second nature to you, and soon, you'll be making things like a pro.

From this point forward, we'll assume you have a grasp of the basics, and we'll forge ahead with the finer topics. Also see Chapter 2, "Robot Tool Crib," for an in-depth review of practical tools used in the constructions of robots.

General Drilling Tips

Regardless of the material (wood, plastic, or metal), keep the following points in mind.

Bit-Depth Control Various methods can be employed to control the depth of the hole in thicker materials. This is needed when you want to drill a hole but don't want it to go all the way through the material. For example, you would want a limited-depth hole in the mandible (jaw) of an android robot for setting a hinge pin. A pin is used as a hinge, and because the holes don't go all the way through the skull and jaw parts, there is no need for fancy clips or other fasteners.

A down-and-dirty method is to wrap masking tape around the shaft of the drill bit. Place the edge of the tape to mark the depth of the hole you desire. A more surefire method is to use a drill collar. This metal collar uses a set screw to firmly attach to the bit. You are assured of an accurate depth because the collar will not allow the bit to go any deeper.

Another type of depth-limited hole is the counterbore, where the drill hole is in different diameters. Counterbore holes (see Figure 5-9) can be created using either of the previous depth-control techniques and by drilling a second, less deep hole with a larger bit. Additionally, specialty counterbore bits are available in several popular sizes. Counterbore holes are commonly used in wood and thicker plastics for countersinking fasteners.

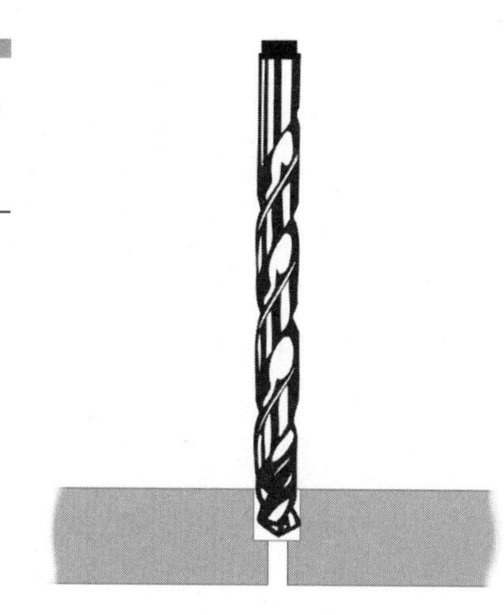

Figure 5-9 An example of a counterbore hole

Aligned Holes Drilling with any handheld tool will naturally produce a certain amount of error. Even the most skilled worker cannot always drill a hole that is at exact right angles to the surface of the material. When precisely aligned holes are a must, use a drill press or drill-alignment jig. The latter can be found at better hardware and tool stores that stock specialized carpentry accessories.

Work Hold-Down and Clamping When drilling with power tools, care must be exercised to hold the part in a clamp or vise. Without appropriate hold-down methods, the drill may cause the material to spin with the bit. Even a piece of wood, turning at a few thousand rpm, can cause serious injury. I once almost lost a finger (or two) when I tried to take a shortcut and manually hold a small piece of metal I was drilling. The bit caught in the metal, and it was ripped from my hand. Luckily, I escaped serious injury, but I decided never to play with fate again.

Hold-down clamps and vises come in various shapes, sizes, and styles. There is no one type that works for every occasion. Spring-loaded clamps (they look like giant tweezers) are useful for very small parts, while C-clamps are handy for larger chunks of material. A vise is required when drilling small parts of any type with a drill press.

Determining Drill Speed Most drill motors lack a means to directly measure the speed of the tool, so the rpm should be considered an approximation only. Go by the sound of the tool. If you know the full speed of your particular tool is 6500 rpm, you can recognize approximate half and quarter speeds by listening to the sound of the motor.

Tips for Drilling

Table 5-3 lists some handy tips for drilling wood, metal, and plastic.

General Sawing Tips

Wood, metal, and plastic can be cut using hand-operated or power tools. For all but the lightest materials, however, you will find that power tools make short work of the job.

For hand tools, the practical choices are as follows:

- For wood, a *backsaw*; for metal and plastic, a *hacksaw*. The backsaw will need periodic resharpening to keep it in top form. For the hacksaw, you need only replace the blades when they become dull.
- A *coping saw* allows you to cut corners with a tight radius in wood, metal, and plastic. A coping saw is similar to the hacksaw except the blades are smaller. Replace the blade when it's dull.
- A *razor saw* is used with thin woods and plastics. Its shape is like that of a backsaw, but much smaller. You can find razor saws at the hobby store.

For power tools, the practical choices are as follows:

- A *jigsaw* is used for cutting wood.
- *Circular saws* and table saws are useful for cutting long, straight cuts in wood and metal. Be sure to use the proper blade, or else damage to the material and blade could result.
- A *bandsaw* can be useful for cutting small shapes in wood, metal, or plastic. The bandsaw is useful when cutting plastic because the long length of the blade (usually 40" or more) stays relatively cool as it goes through the machine. This reduces remelting of the plastic. The better band saws have a speed control.
- A *circular miter saw* (see Figure 5-10) is useful when cutting aluminum channel and bar stock. The alternative is a hacksaw, a miter block, and sweat.
- A *scroll saw* can cut very intricate shapes in wood, metal, and some plastics. The saw cuts by reciprocating action, so the blade can become quite warm when used with metal and plastic. Use a low cutting speed with both metal and plastic to reduce heating.

Table 5-3 *Tips for drilling*

	Wood	Metal	Plastic
General tips	Wood is readily drilled using a motorized drill, either handheld or drill press. Speed is dependent on the size of the bit and the density of the wood. The following are general speed recommendations (all are in rpm). Larger than 1/8": 2100 1/8" to 1/16": 4500 Smaller than 1/16": 6200	Metals should be drilled using a motorized drill. Small parts are more readily drilled using a drill press. The following are general speed recommendations for aluminum and other soft metals (all are in rpm). For harder metals, reduce speed by 50 to 75%. Larger than 1/8": 1000 1/8" to 1/16": 1500 Smaller than 1/16": 2000	For soft plastics (PVC), the speed settings are the same as for wood. For harder plastics (acrylic, polycarbonate), reduce drill speed by 30 to 40%.
Bits tips	Wood bits should be ground to 118°. For cutting all but very dense hardwoods (oak), standard carbon twist drills are more than adequate.	For most metals, bits should be ground to 118°. For harder metals, use 135° (split point) bits. For longer life, consider titanium- and cobalt-coated bits.	Use wood bits for soft plastics. For hard plastics, use a pointed bit designed for acrylic, polycarbonate, and similar plastics.
Cooling tips	Air cooling is sufficient. If wood is very hard and thick, pause every 30 seconds to allow the bit to cool down.	Use cutting oil for heavy-gauge metals. Always avoid excessive heat, as this will dull the bit.	Air cooling is sufficient, but if plastic remelts into the hole, consider using a small amount of water.

Figure 5-10
A circular miter saw

NOTE: *There are plenty of other saw types, but they aren't as common. If you have a preference for one type of tool over another, by all means use it. I prefer using a radial arm saw over a table saw, for example, but I recognize radial arm saws are not everyday finds in most garage shops.*

Of all the tools in the shop, saws are the most dangerous—even hand saws. I need not remind you to observe all safety precautions. Never remove a safety device or guard. Work in a well-lit, unobstructed area. *Use eye and ear protection when the saw is in operation.*

Tips for Sawing

Table 5-4 lists some handy tips for sawing wood, metal, and plastic.

Additional Methods for Cutting Materials

Although a saw is the most common means of cutting materials, there are other methods as well. Select the method based on the material you are cutting and on the demands of the job.

- Very thin (less than 1/8"), hard plastics can be cut by scoring then with a sharp utility knife. Place the score over a 1/4" dowel, and apply even pressure on both sides to snap apart the material.

Table 5-4 *Tips for sawing*

	Wood	Metal	Plastic
General tips	For a hacksaw, band saw, or scroll saw, use a medium pitch blade. For motorized tools, set at highest speed.	For motorized tools, reduce speed to 25 to 50%.	Set speed to 50 to 70%.
Blade tips	Match the blade with the thickness and grain of the material. Circular saw blades are often classified by their use (such as crosscut); use this as a guide.	As a general rule, three to five teeth should engage the metal. Use an abrasive cut-off tool for heavy-gauge ferrous metals.	Circular saw: If possible, use a nonmelt blade made for plastics; if not available, a high-quality plywood blade will suffice. The wider the kerf in relation to the thickness of the blade, the better (this avoids remelting). Hack saw, band saw, or scroll saw: Use an intermediate pitch blade (18 to 24 *teeth per inch* [tpi]) or use a wide kerf blade.
Cooling tips	Air cooling is sufficient.	Use cutting oil or wax for heavy-gauge metals.	Air cooling is usually sufficient. If remelting occurs, direct 50 to 75 *pounds per square inch* (PSI) air from a compressor over the cutting area.

- Thin (to about 20 gauge) metal can be cut with hand or air snips. Pneumatic air snips make the work go much faster. Manual snips are available for straight cuts, left-turning cuts, and right-turning cuts.
- Thinner gauge metal can also be cut using a nibbler (it does what its name implies) and cutout dies. Dies are available in a variety of sizes and shapes. Most common are circles with diameters from about 1/2" to over 2", and cut-out shapes for 9-, 15-, and 25-pin D-style computer cable connectors. To use the die, you merely drill a pilot hole, assemble the two halves of the die, and tighten with a wrench. The die punches through the metal as you tighten it.
- For higher gauges of metal, and for long, straight cuts in thinner gauges, cut using a bench shear.
- Foam-based materials that are not laminated (Styrofoam) can be cut using a hot wire. Hot wire kits are available at most craft stores.
- Heavy plate steel and pipe can be cut using a gas torch.

Metal Brakes and Shears

When working with sheet metal, nothing beats the metal brake and shear. These tools are either bench or floor mounted and allow you to work with large pieces of sheet metal. The brake is used to bend metal; the shear cuts it.

Industrial brakes and shears are typically separate tools, but combination brake/shears are ideal for small shops. They save space and money. Most of the affordable home-shop break/shear combinations are limited to working with sheets under about 30" wide.

Brakes and shears are limited in the gauge of metal they can accept. For the smaller tools, capacities range to about 20 or 22 gauge for mild steel; larger industrial tools can accept up to 16 or 18 gauge metal.

Using Portable Power Tools

Your work in the robot shop will go much faster if you use power tools. This applies even if you're constructing your 'bot out of wood. The use of power tools is fairly straightforward, and the subject needs little additional instruction beyond what you'll get in the instruction manual and in a good shop tools book. Here we compile some of the important issues to be aware of.

Motor

- A motor that can operate with both DC and AC voltages is termed a universal-type motor. It is best used at the specified voltage, or overheating will result. Check the tool and avoid the use of long extension cords.
- Current is fed through stationary brushes to the commutator and then to the armature, which drives the tool. When the brushes fail to make good contact with the commutator, proper current will not reach the armature, and the tool may either fail or may not deliver full power. Brushes wear down by constant friction against the commutator.
- For tools with replaceable brushes, inspect them periodically (every year for a tool that sees moderate use; more frequently for a tool used more often). This is done by removing the brush caps, and then removing the brushes and springs. Replace the brushes when they are worn to less than 1/4", or as specified by the manufacturer.
- Brushes may be cleaned with turpentine, and many can be reground to a better shape (to match the curvature of the commutator).
- Sparking in universal motors is fairly common, but excessive sparking may indicate a damaged motor.
- The grounding lug on AC-operated tools should *never* be defeated.
- When working outside, use an AC outlet equipped with a ground-fault interrupter circuit.

Maintenance

- Many power tools include a fan, attached to the shaft, for cooling the motor. The fan should be kept clean and free of obstruction. Clean if dirty. Do not use the tool if metal bits have become caught in the cooling vents.
- Many tools should be periodically lubricated, as indicated by the manufacturer. This is especially true of reciprocating saws. Gear-driven tools (such as heavy-duty power drills or worm-drive circular saws) should be lubricated with the appropriate grease or high-viscosity oil. (On units with sealed drive systems, relubrication is not generally needed unless the tool sees heavy use.)

General

- When using an extension cord, be sure it is rated for the current being passed through it. Otherwise, the cord may overheat and could cause a fire. As a general rule, medium-duty tools (8 to 10 amps) can be used with 14 gauge extension cords, but keep the cord under 50 feet. Higher amperages, and longer lengths, require larger gauge (12, 10, and even 8 gauge) extension cords.

- Keep the exterior of the tool clean by wiping it off with a dry cloth. Grime can be removed with a slightly damp cloth. Never apply water or any other liquid directly to the tool.

Horsepower

- Horsepower for (AC-operated) tools is calculated as follows:

$$\frac{\text{volts} \times \text{amps}}{746} = \text{hp}$$

Example:

$$\frac{115 \times 10}{746} = 1.54 \text{ hp}$$

- Don't get caught up with the horsepower rating of your tools. More important is how well they do the job, and raw horsepower is only one factor. A saw with high horsepower may still do a poor job if the blade is dull or mounted improperly.

Using Pneumatic Tools

Electric power tools are by far the most common variety found in any home shop. Yet another kind of power tool is the pneumatic (air) type, which is driven by a blast of high-pressure air. Common pneumatic power tools include the drill (see Figure 5-11; standard and reversible models are available), air hammers, metal snips (*very* handy!), and the power wrench. Pneumatic tools offer several advantages:

- The tools are generally less expensive than their electrified cousins, because their motor is a fairly simple air pump. Offsetting this, however, is that you need an air compressor to supply the air.
- Air tools tend to be lighter, again because they lack a bulky AC or DC motor.
- They're safer to use outdoors, because there's no electricity to the tool. You can keep the compressor indoors and connect the compressor and tool via a hose.
- There are no brushes to replace.
- Properly cared for, air tools can last a lifetime (your lifetime, that is).

So now that you're sold on pneumatic tools, here are some basic tips for using them.

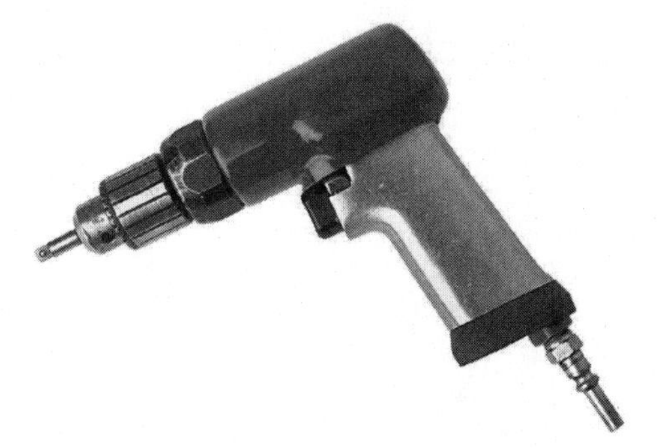

Figure 5-11 Pneumatic tools like this one offer a number of cost benefits over electrically powered tools.

Water and Oil

Water destroys air tools. Oiling your tools every day that you use them will greatly prolong their life and help dispel trapped water. Even if you have an air dryer attached to your compressor, be sure to add oil to the tool (simply squirt a dab or two into the air intake) the first time you use it that day.

You may also wish to invest in an inline oiler, which adds oiled air to the tool as you use it. The oiler connects to the hose line running from the compressor to the tool.

Air Pressure and Volume

Pneumatic tools require both air pressure and volume to work. Some air tools, such as drills, don't need a great volume of air, but they need high pressure. Be sure your compressor can deliver both the pressure and the volume needed by your tools. The specifications that come with each tool will tell you what it needs. As a general rule, a compressor that can constantly deliver 85 to 90 psi, and 4 to 8 *cubic feet per minute* (cfm) will properly power the average collection of pneumatic tools.

NOTE: *Some air tools, especially saws and some grinders, may require in excess of 100 psi and more than 15 to 18 cfm.*

Hoses and Connectors

Keep the air hose as short as possible. The longer the hose, the less pressure makes it to the tool. A maximum length of 25 feet is ideal. I prefer the use of non-coiled hoses, because the straight hoses tend to provide higher air pressures. The coiled hoses are more convenient, however.

Equip all your hoses and pneumatic tools with quick-disconnect couplers, such as that in Figure 5-12. This allows you to change the tool simply by popping it off the coupler. Otherwise, you must use a wrench to loosen and tighten the connection each time. The better tools come with quick-disconnect couplers; if yours lacks a coupler, they're relatively inexpensive when purchased separately.

Ear and Eye Protection

One final word: Wear ear plugs (available at any drug store) when using air tools. Your eardrums will thank you for it. And of course, always wear eye protection when using any power tool.

A Look at Desktop Mills and Lathes

Model makers have used so-called *desktop* mills and lathes for years. These tools are similar in features and function to the behemoths you see at large machine shops and on manufacturing floors but are scaled down in both size and price—an ideal mix for amateur robot builders.

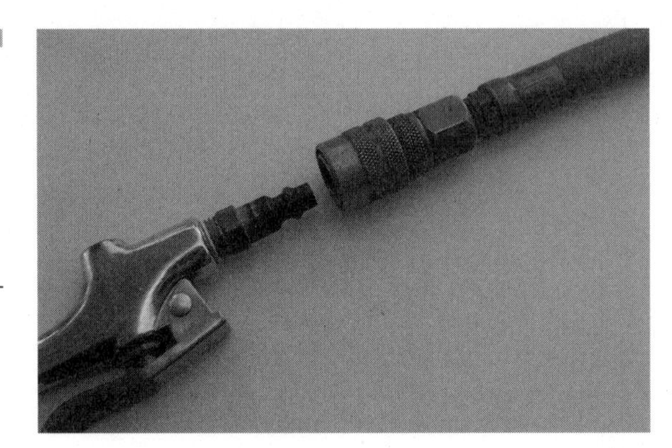

Figure 5-12 Quick-disconnect couplers allow you to rapidly change air tools.

First, some definitions:

- A *mill* is like a vertical drill press. Instead of a cutting bit that just goes up and down, the work piece on a mill can be moved horizontally and laterally. This allows the mill to produce complex shapes instead of just holes.
- A *lathe* is used to rotate a part against a cutting tool. It is typically used to contour round or cylindrical shapes—threads on a rod, for example. The material turns on a horizontal bed; the cutting tool is brought up against the material.

A third type of machine, the *computer numerically controlled* (CNC) router, is a high-speed cutting tool, like a wood router, that is attached to a mechanism that moves the router in the X-, Y-, and Z-axes. This movement is managed by a computer. More about this tool will be discussed later in the chapter. Figures 5-13 and 5-14 show a desktop mill and desktop lathe, respectively.

Makers of Desktop Machinery

A number of makers of desktop mills, lathes, and CNC routers are on the market today. Among the more common manufacturers are

- Cool Tools
- FlashCut CNC

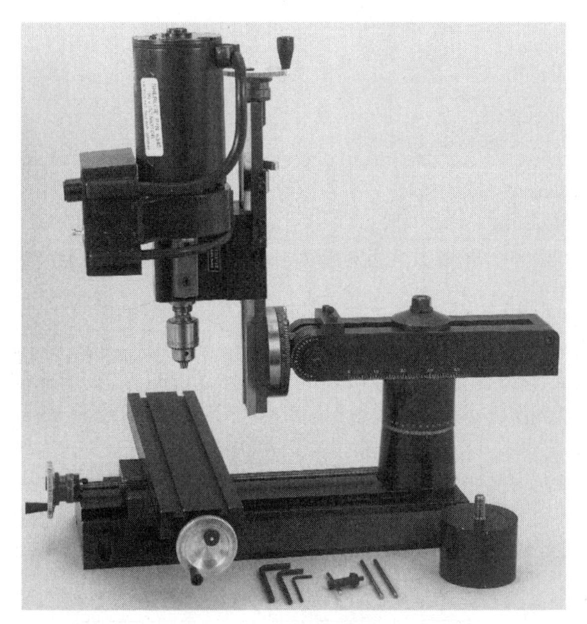

Figure 5-13 A desktop mill can be used to create miniature parts for your robot.

Figure 5-14 Desktop lathes can be used to create rod stock, shafts, specialty fasteners, and more.

- ISMG (International Sales and Marketing Group)
- Techno-Isel
- Liberty Enterprises
- Many/EasyCut
- MAXNC
- MicroKinetics
- Minicraft
- Minitech
- Sherline
- Super-Tech
- TAIG

The prices and features from these makers vary. If you're interested in acquiring a desktop mill, lathe, or CNC router, you're well advised to get information on as many of them as possible. The typical starting price for the better-made tool is $500, with many in the $2,000+ range, so consider researching part of your investment.

Not all desktop tools are created equally. Some are designed for garage-shop tinkerers on a budget. They're fine for working with lightweight materials like soft plastics and thin woods, but don't think you can use them to produce highly accurate complex shapes from stainless steel! Price goes up based on accuracy, power, and size, so plan your purchase accordingly. If you need to work with pieces up to 20", don't settle for a machine with a maximum cutting size of just 18".

One way to save money on a desktop mill or lathe is to purchase it used. The better machines fetch good prices on eBay and other online auctions, but you may have good luck snagging a steal simply by going to garage sales and checking the local newspaper classified ads.

CNC Routers

A CNC router is inherently a computer-controlled device. Mills and lathes can be completely manual affairs, or they too can be hooked up to a computer. With most models, you can purchase a manual desktop mill and lathe today, and sometime down the road retrofit it for computer control.

For the robotics hobbyist with a bit of extra spending cash, a *CNC router* will have you building all sorts of 'bots in record time. A CNC router combines a high-speed motorized cutting tool (the router) with computer-controlled movement. This movement is controllable in three planes: X, Y, and Z.

To use, you first secure a piece of wood, plastic, metal, foam, or other material onto the base of the device. You then program the moves the router will take over the material. For example, if you're cutting out a shape, the router will first move downward (the Z-axis) to drill into the material, and then move back and forth (the X- and Y-axes) to cut out the shape.

Depending on the routing tool used in the machine, you might also be able to mill parts out of softer materials, such as plastics and soft woods. (For metals and other hard materials, a mill is the better choice, as it has more cutting power.) When milling on a CNC router, the Z-axis of the router is varied ever so slightly, in order to produce a 3D cut surface.

Although CNC routers are simple in principle, they are not cheap in cost. The typical CNC router package for hobbyist use costs $2,000 or more. Prices go up from there, with $5,000 being typical for an entry-level professional model.

The design of the typical CNC router is the *gantry*, like the one in Figure 5-15. The gantry slides back and forth, and is the X-axis. Attached to the gantry is a Z-axis plate, which moves up and down; the router tool is physically attached to this. Depending on the design of the machine, the Y-axis is produced by either moving the entire gantry or by moving the work piece itself (see Figure 5-16). The latter is probably more common in the smaller desktop CNC machines—the ones most suitable, in size and price, for the amateur robot builder.

The *travel distances* of the three axes define the maximum size of material you can work with. A small CNC router may be limited to a 12" × 12" piece of material, and with a maximum thickness of 2" to 3". When comparing CNC routers, check the extents of the X-, Y-, and Z-axes and be sure they will be adequate for your needs.

You'll want to verify that these dimensions are travel extents (the tool actually travels this distance to cut) and not merely the maximum dimensions of the material you can fit into the machine. A given CNC router may be able to accept material up to 12" × 12", but may only be able to cut out a shape of 8" × 11".

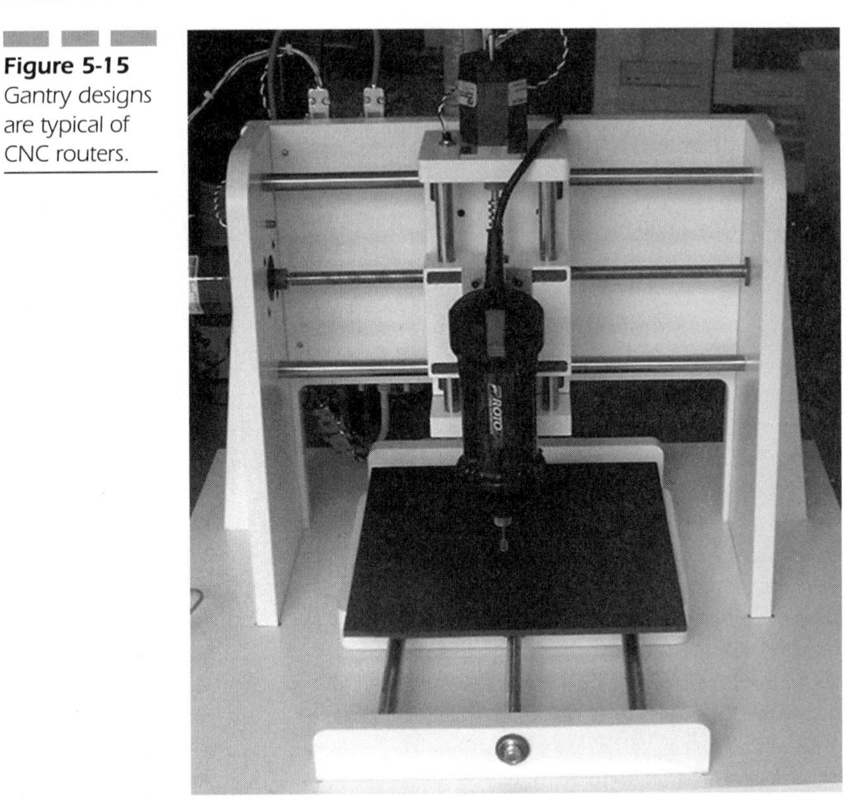

Figure 5-15
Gantry designs are typical of CNC routers.

Figure 5-16
The router cuts and mills material by moving the tool, the material, or both.

Figure 5-17 Stepper motors provide exact movement of the router for precision cuts.

As depicted in Figure 5-17, the movement of the three axes of the CNC router is performed by a *stepper motor*, or the less common servo motor, which adds considerably to the cost of the machine. The stepper motor drives the mechanics of the CNC router via an acme screw, ball screw, trapezoidal lead screw, rack and pinion gear, belt, or chain. I'll let the manufacturers tout the pros of their specific systems, but in the end, you'll want to ensure your machine has the *repeatable accuracy* you need for your work. Any CNC machine with a repeatable accuracy of less than 0.010" is not worth your investment.

Note the term *repeatable* above. Some CNC router vendors list the positioning accuracy of the stepper (or servo) motor. This is not the same as the repeatability of the cutting tool, which takes into consideration the flatness of the work table, the type of drive mechanism, the effects of backlash as the tool moves back and forth, and other variables.

One simple way to test the accuracy of a CNC router is to replace the router tool with a fine-tipped felt pen. Securely tape a piece of paper to the work table and have the router draw a shape—such as the figure of a dog or person—onto the paper. Do it twice. Carefully examine the drawing: You should not see double traces anywhere. If you do (it will likely be at the corners or intersections of lines), that router lacks sufficient accuracy.

When purchasing a CNC router, consider the *software* you will use with the machine. Many commercially made CNC routers come with software; others don't. CNC software can cost several hundred, to several thousand, dollars; if your machine lacks software, be sure to add this cost to the final price. (Note: Most CNC routers can be used with software from a variety of vendors, but it's still a good idea to make sure yours doesn't use some proprietary control technique that limits your choices.)

Finally, if the cost of a ready-made CNC router is too rich for your blood, you might want to consider making your own. It's not *quite* as easy as some Web sites and magazine ads make it out to be, but you can save 40 to 60 percent.

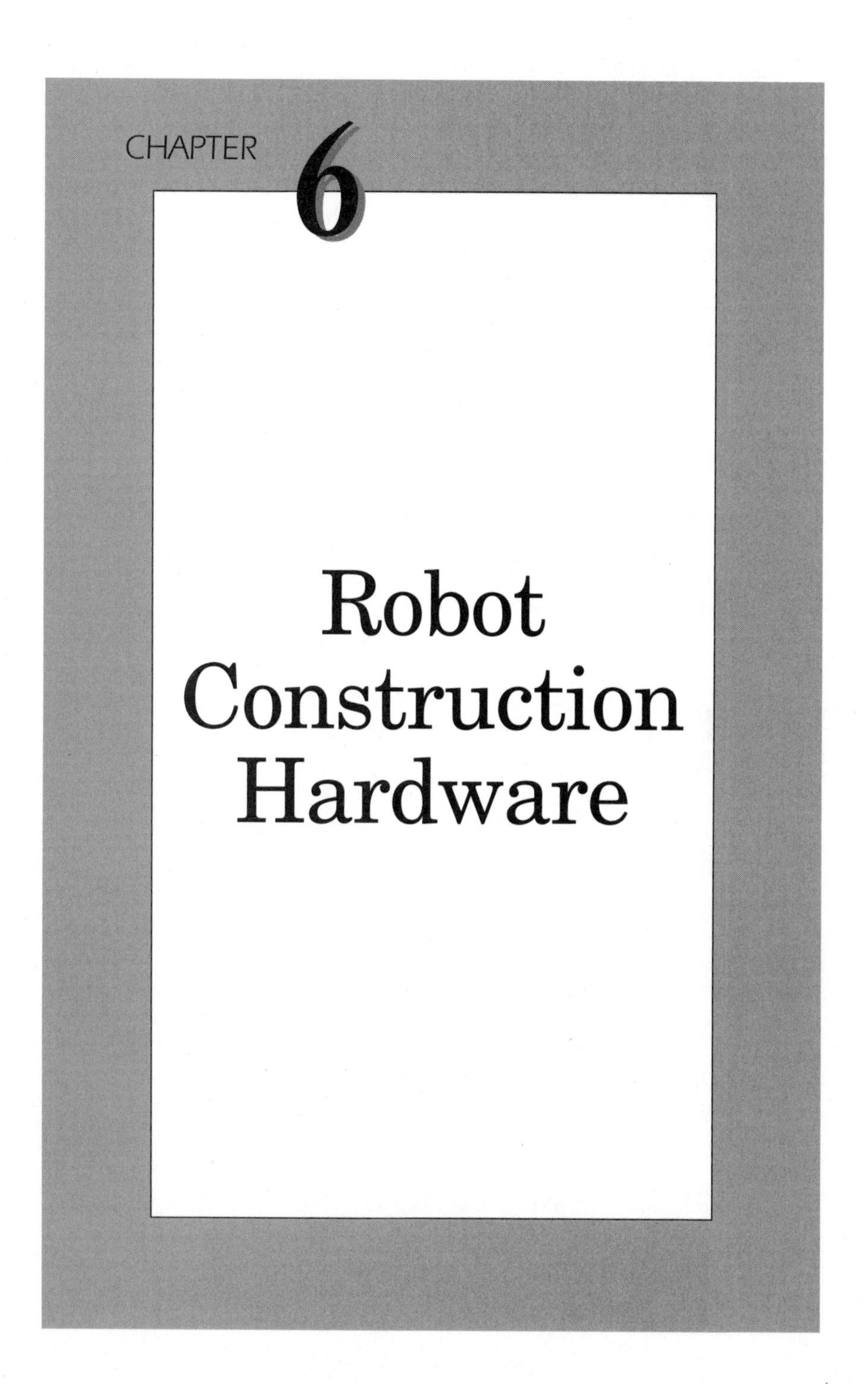

CHAPTER 6

Robot Construction Hardware

A robot is an amalgam of parts, small and large, important and seemingly inconsequential. We tend to attend to the big stuff: motors, wheels, batteries, and bases. But it's easy to forget how the robot is put together, and for most machines, that's with hardware.

It's easy to spot those robots given short shrift by their makers in terms of hardware. They tend to fall apart, even during normal handling. More often than not, they act erratically, never repeating the same sequence twice. With the right hardware and construction techniques, your robots can last longer and work better.

Hardware is more than nuts and bolts; it's brackets and gussets and aluminum tubing and little plastic rollers and nylon spacers and tens of thousands of other items. Many of these items can be purchased at local retailers, including hardware, home improvement, hobby, and craft outlets. Thanks to the magic of mail order and the Internet, what you can't find locally, you can find globally. Many online retailers provide a virtual catalog that you can browse to find what you need.

Making the best use of hardware is not only knowing what's available, but imagining how it is used to construct mobile autonomous machines. In this chapter, we'll look at hardware that's proven to be useful in building robots. You may already have some of this hardware in your shop.

Fasteners

Fasteners are the most elementary of all hardware. There are dozens of fastener types, but we'll concentrate on those that are most practical for amateur robotics. These are machine screws, nuts, washers and all-thread rod (see Figure 6-1).

- *Machine screws* are designed for fastening together the parts of machinery, hence the name. Unlike a wood or metal screw, machine screws do not have a pointed end. The machine screw is designed to be secured into a nut or other threaded retainer. If the material being joined is threaded, the screw can be secured directly. Machine screws are sometimes referred to as *bolts*, and when a nut (see the following) is included, they're known as *stove bolts*. I like to limit the term bolt to a machine screw larger than 1/4" in diameter, though this is by no means universal. You're free to use any terminology you like.
- *Nuts* are used with machine screws. The most common is the hex nut, so called because the nut has six sides. The nut is fastened using a wrench, pliers, or hex nut driver. Although hex nuts are the most common and the cheapest, there are other types, including *square nuts* and *T-nuts* (also called blind nuts). These are most commonly used with wood and soft plastics. Also

Figure 6-1 Machine screws, nuts, washers, and sections of all-thread rod are among the most common and practical fasteners used in robotics.

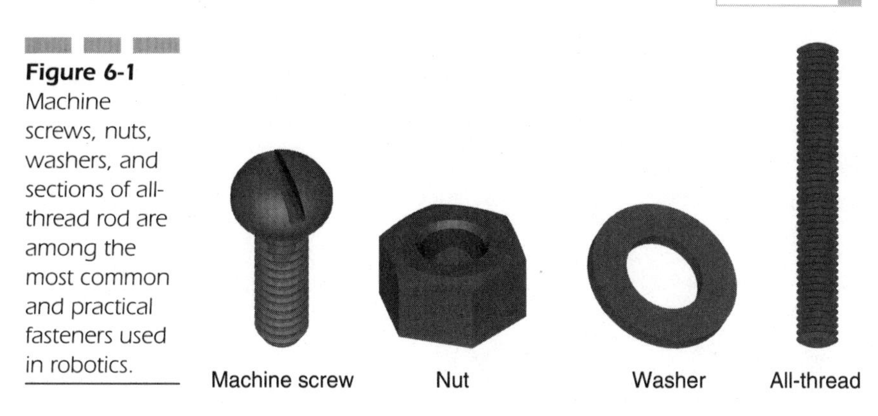

handy in many robotics applications are *locking nuts*, which are like standard hex nuts but with a nylon plastic insert. The nylon helps prevent the screw from working itself loose.

- *Washers* act to spread out the compression force of a screw head or nut. Under load and without a washer, damage may occur because of the small contact area of a machine screw head or nut. The washer doubles, or even triples, the contact surface area, spreading out the force. Washers are available in diameters to complement the size of the machine screw (or bolt for larger sizes). Variations on the washer theme include *tooth washers* (internal or external) and *split lock washers,* which provide a locking action to help prevent the fastener from coming loose. To be effective, the locking washer must be correctly matched with the size of fastener with which it is used.
- *All-thread rod* is rod with common machine screw sizes and threads. In fact, it's commonly used when a specific length of machine screw is not available. All-thread is sold in 1" to 6" lengths. Apart from making your own custom-length screws, all-thread is good for making shafts and linear motion actuators.

Name Mix and Match

Making fasteners is among the oldest job descriptions of the industrial revolution. They've been doing this for hundreds of years, and naturally, the names of things can vary, depending on who you talk to. In this chapter, and throughout this book, I attempt to refer to the various types of fasteners by their most common names, as used in the United States (where I'm from). But this doesn't mean other names for the same fasteners aren't equally valid.

I can't even start to list all the variations, and it's not the point of this book to be a tell-all compendium on fasteners. But here are a few of the different names for the same thing you may encounter:

- Acorn nut = cap nut
- Step bolt = carriage bolt = square shoulder bolt
- Philips head = cross head
- T-nut = blind nut
- Hangar bolt = threaded stud

The list could go on and on. When searching for the right fasteners, be on the lookout for products that have different names than you expect. Web sites that provide illustrations of their offerings make it easier to locate exactly what you want, even if they use a different name than that to which you're accustomed.

Fastener Sizes

Fasteners are available in common sizes, either in metric or standard.

NOTE: *Standard is often referred to as English, American, fractional, inch, unified, UN, UNR, and several other descriptors. The English in the U.K. are now firmly in the metric camp and standard-size fasteners are used by more than just the Americans. Fractional and inch sound confusing, and what could be unified about a screw thread system primarily used in only one country of the world? So, for the lack of a single perfect word, throughout this book I use either standard or SAE, the latter of which stands for Society of Automotive Engineers.*

Standard-size fasteners are denoted by both their diameter and the number of threads per inch. For example, a fastener with a thread size of 6-32 (also shown as 6/32) has a diameter referred to as #6, with 32 threads per inch (also called *pitch*). Diameters under 1/4" are indicated as a # (number) size; diameters 1/4" and larger can be denoted by number, but are more commonly indicated as a fractional measurement—3/8", 7/16", and so on.

The pitch can be either *coarse* or *fine* for standard fasteners. Therefore, not all #6 fasteners have 32 threads to the inch. Table 6-1 shows screw sizes from #1 to #24. As indicated, note that sizes #14 and above are more commonly referred to using a fractional measuring system.

Table 6-2 shows the differences in threads per inch for screw sizes #4 to 1", in both *unified course* (UNC) and *unified fine* (UNF) pitches. Coarse threads are by far the most common at hardware stores, though #10 screws are routinely avail-

Table 6-1 Screw sizes at a glance

Screw Size	Inch Fraction	Decimal	Closest Metric Equivalent (mm)
#1	1/16	0.0625	1.58750
#2	5/64	0.078125	1.98437
#3	3/32	0.09375	2.38125
#4	7/64	0.109375	2.77812
#5	1/8	0.125	3.17500
#6	9/64	0.140625	3.57187
#8	5/32	0.15625	3.96875
#9	11/64	0.171875	4.36562
#10	3/16	0.1875	4.76250
#11	13/64	0.203125	5.15937
#12	7/32	0.21875	5.55625
#13	15/64	0.234375	5.95312
#14	1/4	0.250	6.35000
#16	17/64	0.265625	6.74687
#18	19/64	0.296875	7.54062
#20	5/16	0.3125	7.93750
#24	3/8	0.375	9.52500

able either coarse or fine. This can actually be a problem, because it's easy to buy 10-32 nuts and not notice you have 10-24 machine screws. Bummer!

Metric Sizes Metric fasteners don't use the same sizing nomenclature as their standard cousins. Screw sizes and pitches are defined by diameter, the thread pitch (number of threads per millimeter for metric), followed by length—all in millimeters. For example: M2-0.40 × 5 mm means the screw is 2 mm in diameter, has a pitch of 0.40 threads per millimeter, and has a length of 5 mm. Note that metric screws use a normalized thread pitch—none of this course or fine stuff—so the pitch may be omitted: M2 × 5 mm.

Table 6-2 Standard threads at a glance

UNC (coarse)	UNF (fine)
4-40	4-48
5-40	5-48
6-32	6-40
8-32	8-36
10-24	10-32
1/4" x 20	1/4" x 28
5/16" x 18	5/16" x 24
3/8" x 16	3/8" x 24
7/16" x 14	7/16" x 20
1/2" x 13	1/2" x 20
5/8" x 11	5/8" x 18
3/4" x 10	3/4" x 16
7/8" x 9	7/8" x 14
1" x 8	1" x 14

Table 6-3 compares metric screw sizes. Table 6-4 details common metric thread pitches. Table 6-5 shows the nearest equivalent standard screw size, for the purpose of substituting metric screws when plans or instructions call for standard.

Anatomy of a Machine Screw

Machine screws have the following characteristics, shown in Figure 6-2:

- **Head** The head is used to drive the screw. Actually, not all screws have heads, but most do. Set screws are a good example of screws that don't always have heads.
- **Shoulder** The shoulder is the area of the screw right under the head. For most machine screws, the shoulder is threaded, but for other types of screw (like the wood screw or carriage bolt), the shoulder is plain or squared.

Table 6-3 Metric screw sizes

Diameter	mm	Inch	Diameter	mm	Inch
M1	1	0.0393	M20	20	0.7874
M1.1	1.1	0.0433	M22	22	0.8661
M1.2	1.2	0.0472	M24	24	0.9448
M1.4	1.4	0.0551	M27	27	1.0629
M.17	1.7	0.0669	M30	30	1.181
M1.8	1.8	0.0708	M33	33	1.299
M2	2	0.0787	M36	36	1.417
M2.2	2.2	0.0866	M39	39	1.535
M2.3	2.3	0.0905	M42	42	1.654
M2.5	2.5	0.0984	M45	45	1.772
M3	3	0.1181	M48	48	1.890
M3.5	3.5	0.1378	M52	52	2.047
M4	4	0.1574	M56	56	2.205
M4.5	4.5	0.1771	M60	60	2.362
M5	5	0.1968	M64	64	2.520
M6	6	0.2362	M68	68	2.677
M7	7	0.2755	M72	72	2.835
M8	8	0.3149	M76	76	2.992
M10	10	0.3937	M80	80	3.150
M12	12	0.4724	M85	85	3.346
M14	14	0.5511	M90	90	3.543
M16	16	0.6299	M95	95	3.740
M18	18	0.7086	M100	100	3.937

- **Shank** The shank is the threaded portion of the screw. The shoulder and shank together determine the length of the screw. The diameter of the shank determines the size of the screw.
- **Thread** All screws are, by their nature, threaded. The threads can be left or right; most screws use right-handed threads. The number of threads per inch or millimeter determines the pitch of the screw.

Table 6-4 Common metric coarse thread pitches

Size	Pitch	Size	Pitch
M2	0.40	M24	3.00
M3	0.50	M27	3.00
M4	0.70	M30	3.50
M5	0.80	M33	3.50
M6	1.00	M36	4.00
M8	1.25	M39	4.00
M10	1.50	M42	4.50
M12	1.75	M45	4.50
M14	2.00	M48	5.00
M16	2.00	M52	5.00
M18	2.50	M56	5.50
M20	2.50	M64	6.00
M22	2.50		

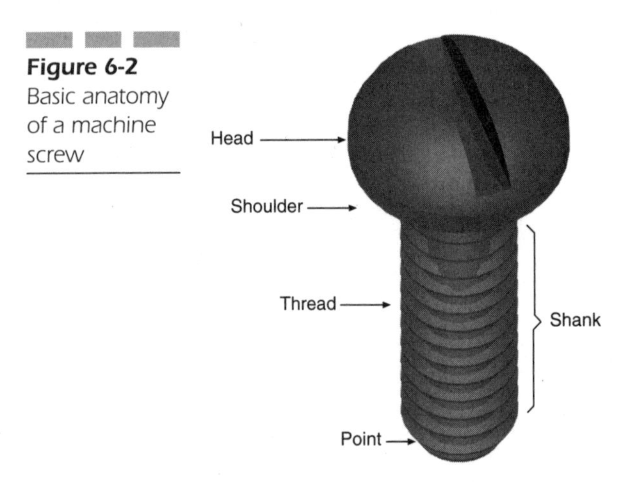

Figure 6-2 Basic anatomy of a machine screw

Table 6-5 Nearest equivalent standard screw size

Metric	Standard
M1.7	#1
M1.8	
M2	#2
M2.2	
M2.3	#3
M2.5	
M3	#4
M3.5	#6
M4	#8
M4.5	#9
M5	#11
M6	#13
M7	#16
M8	#20
M10	#22

- **Point** The typical machine screw has a blunt (or *die*) end, but other screw types have cone, cup, or taper points. Self-drilling screws may have a cone or *pinch* point, along with flutes that dig into the material. Set screws are usually pointed or cupped, depending on the application.

Fastener Head Styles

When buying machine screws, you have a choice of a variety of heads and drivers. The head greatly contributes to the amount of torque that can be applied to the screw when tightening it. Additionally, certain head styles are designed to have a lower profile, meaning they stick out less than the others.

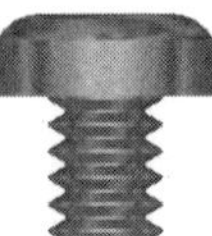

Figure 6-3

Pan

A pan head screw is a good general-purpose fastener. However, the head is fairly shallow, so it provides less grip for the driver.

Figure 6-4

Round

A round head screw has a taller head that protrudes more than a pan head, so it provides greater depth for the driver. It is good for higher torque applications when it doesn't matter if the head sticks out.

Figure 6-5

Flat (or Countersunk)

A flat head screw is used when the head must be flush with the material's surface. These require a countrsunk hole.

Figure 6-6

Oval

An oval head screw is often used as a substitute for a flat head screw when the head requires extra depth. The top of the oval head is semirounded.

Figure 6-7

Button

A button head screw is much like a round head screw, but is commonly used with specialty drives (see the next section on drive types).

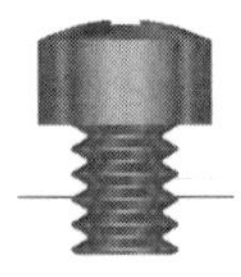

Figure 6-8

Fillister

A fillister head screw has an extra deep head for very high torque. The top of the head is rounded.

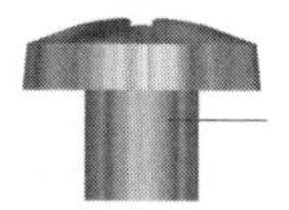

Figure 6-9

Binding

A cross between pan and fillister, the binding head provides a wide contact area and is ideally suited for making electrical connections to solid and stranded wire.

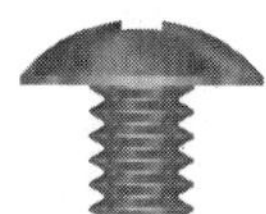

Figure 6-10

Truss Head

A truss head is a very large head and is often used with thin sheet metal to provide the largest possible surface area without the use of a washer. The head covers holes that may have been drilled oversize.

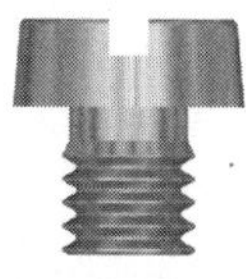

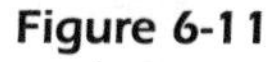

Figure 6-11

Cheese

A cheese head is similar to a fillister, except the top of the head is flat instead of rounded.

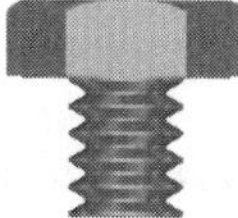

Figure 6-12

Hex Bolt

A hex bolt uses no slot and requires a wrench to tighten. Use it for the highest torque applications.

Round, pan, and flat head screws are by far the most common, and they tend to be the least expensive. These are the most common, but there are at least a dozen others. They don't see heavy use in most robotics applications, so I won't list them here.

Other variations increase the available choices:

- *Sems* is a subtype of screw. Sems screws are threaded fasteners that are combined with captivated washers, collars, unique stampings, or other components (captivated means the washer, collar, etc. will not readily come off).
- *Attached washers* are stamped into the metal of the fastener (either the screw or the nut). This saves you from adding a washer of your own.

- *Undercut, self-sealing, cut angle, etc.* are further options. Many of the head styles, particularly flat head (where the cut angle can vary from 70 to over 100 degrees), are available with extra machining for specialized applications. Unless you require these types, there is no need to bother with them as they are more expensive than standard fasteners.
- *Fillister screws* are available with and without a hole drilled in the screw head. The hole is typically used with automated assembly equipment.
- *Thread-forming screws* combine the shank profile of machine screws with the steep helix thread of wood or metal screws. Many thread-forming screws are designed for plastic, expanded foam materials, and composites. They are designed to minimize cracks and breaks due to compression. Care must be taken to ensure the proper fastener for the plastic you are using. Check with the manufacturer or distributor to ensure the angle of the thread profile is suitable for the plastic you are using. You need a distinct thread profile for thermoset plastics (they can't reshape with heat) as you do for thermoplastics (they can reshape with heat).
- *Shoulder screws* have a pan, oval, or other style of head, with a square, rather than round, shoulder. Recall that the shoulder is the part of the screw right under the head.

Fastener Drive Styles

Most machine screws available at the hardware store are slotted for flat-bladed screwdrivers. You may instead wish to use Phillips, Torx, square drive, Pozidriv, or hex head (Allen wrench) screws. All require the proper screwdriver. Slotted screws are cheaper to make, so they cost less. But there's a risk of stripping out the slot if they're overtightened. Specialty drive screws can be tightened and loosened without as much risk of stripping out the head.

Here is an overview of the several commonly available drive types and what makes them unique. For all types, different sizes of drivers are used to accommodate small and large fasteners. In general, larger fasteners need larger drivers. And the larger the driver, the more torque that can be applied for a sure fit.

Which to use? I personally prefer Phillips, as long as the proper size screwdriver is used. A Phillips head screw can generally accept a greater amount of torque than the same size slotted screw can accept. They are particularly well behaved when using motorized screwdrivers. Slotted screws provide some latitude in the size of screwdriver bit used with them, so an exact match between screw and bit is not required.

NOTE: *Remember, the cross-point head of a Phillips screw can be greatly mangled by using a too-small or too-large screwdriver bit, so choose your tools carefully.*

Combination (slotted and Phillips) head screws are handy when the proper tool isn't available. But as combination head screws are more expensive and harder to get, it makes sense to have the proper tool in your toolbox.

When screws must be heavily tightened, hex, Torx, and square drive types are recommended. Do note they are more expensive than either Phillips or slotted screws. Pozidriv screws are among the most expensive and require a drive bit that most hardware and home improvement stores in North America do not carry (Pozidriv is more popular in Europe). You need to purchase a Pozidriv bit from a specialty retailer, via mail order, or from the better-stocked home improvement outlets.

Fastener Materials

The most common metal faster is steel plated with zinc. These resist rust and are quite affordable, even in small quantities. Yet there are many other materials and plating for fasteners, including aluminum, titanium, and even plastic. The following applies to all fastener hardware: screws, nuts, washers, and so on:

- *Zinc-plated steel* is the fastener of choice because of its ready availability and low cost. Typical steel fasteners are made with 1006-1038 carbon steel, which exhibits high tensile strength but can be readily machined. Steel fasteners are magnetic, which can affect certain compass sensors in robots.
- *Stainless steel* offers added strength and a resistance against rusting or corrosion. Stainless steel fasteners do not need to be plated because the material already resists rust and other corrosion. Stainless steel fasteners are commonly available in any of several alloys: 316 stainless steel is nonmagnetic and has extended corrosion resistance; 18-8 stainless steel can be mildly magnetic, exhibits good corrosion resistance, and is competitively priced.
- *Brass* is a softer metal that's most often used for looks. No plating is necessary as brass resists rust. Brass fasteners are naturally nonmagnetic. Note that some brass fasteners are really brass-plated steel, and some steel fasteners are really zinc-plated brass. It's best to look closely on the label to determine what you're getting.

Figure 6-13

Slotted

Made for general fastening and low-torque drive; screwdriver may slip from the slot.

Figure 6-14

Philips

A Philips, or cross-point, drive resists drive slippage, but the head is easily stripped out when using an improperly sized driver.

Figure 6-15

Combination

A combination drive accepts both slotted and Phillips drivers, in case you don't have the right screwdriver handy.

Figure 6-16

Hex, Torx, Pozidriv, and Square

The specific size and type of driver is required, which minimizes stripping. The hex drive is also called Allen, after the company that helped popularize this drive type. Disadvantage: You must have the proper tool to fasten and unfasten.

Figure 6-17

Socket

Combines a fillister or cheese head style with a hex or Torx drive. The rim of the head may also be knurled in order to assist in hand tightening.

- *Aluminum* is used when a metal fastener is desired and weight is a major consideration. It's also the preferred fastener for aluminum structure, as it will not cause corrosion, which can happen if you use steel fasteners with aluminum framing. Threads in aluminum fasteners are more prone to stripping, so avoid using aluminum in high-stress applications.
- *Titanium* offers supreme strength for its weight. Titanium is only modestly heavier than aluminum but has the strength of steel. Unlike those made from steel, titanium fasteners are not magnetic. Some titanium fasteners

are only titanium-plated steel. Note that steel fasteners can be tempered and hardened to improve their strength. As you can imagine, titanium fasteners can be frightfully expensive, but when strength is critical, they're one of the best choices.

- *Monel* is a nickel-copper alloy that exhibits very good corrosion resistance, strength, and machinability. Only the specialists of specialty fastener outlets carry this stuff, and the price tends to be pretty high.
- *Silicon bronze* is made of 95 percent to 98 percent copper, plus silicon for added strength. These fasteners tend to be quite expensive and are used primarily in applications where fumes and gases may corrode metal. Being mostly copper, silicon bronze fasteners are nonmagnetic.
- *Nylon* (or more generically, polyamide plastic) is considerably lighter than steel or brass and is advantageous when weight is a concern. Generally available in two colors: natural (off-white) and black. Natural nylon is far more common and tends to be cheaper. Obviously nonmagnetic, nylon fasteners can also withstand higher temperatures that most other plastic fasteners—up to 300 degrees. One use of nylon fasteners, other than lowering the weight of a robot, is as a low-friction glide. See the entry for Teflon for more detail.
- *Polycarbonate* is a high-impact thermoformable plastic. Polycarbonate fasteners are not common, and they tend to be expensive. They are best used when high electrical insulation is required.
- *Teflon* fasteners are also used for electrical applications. Acorn nuts made of Teflon are also handy as they can be used as low-friction glides on small robots. The robot rolls on two wheels, balanced by these nonrotating Teflon glides.

Then there are various platings and finishes. You've already read about zinc-plated steel. There's also the following:

- *Black oxide finish* can be applied to many base metals, but it mostly is used with steel and stainless steel fasteners. The fastener is black but is not painted (there are also painted fasteners, typically used for home construction; these have little use in robotics). Besides a higher-tech look, black oxide finish resists corrosion and rust better than zinc does.
- *Hot-dipped galvanized* is ideal for applications where water or the whether might cause rust or other corrosion. A lawnmower robot is a good candidate for galvanized fasteners.
- *Colored zinc plating* adds some panache to your robot—green and yellow are common choices.
- *Chrome plating* is added over steel (sometimes brass) for a bright, hardy finish. Great for a hot rod robot with a 458 V8 engine! Seriously, the chrome plating is basically for looks.

- *Nickel plating* is applied to steel for a thicker and longer-lasting coating than is provided by zinc. It's cheaper than chrome plating and provides most of the same luster. Use this if you need to spiff up your combat robot.
- *Anodizing* deposits a thin, often colored, protective film over aluminum fasteners. Common anodizing colors are clear and black, but most any color is possible—reds, blues, greens, you name it.

Customized Finishes

Not every fastener in every size and material is available with every conceivable finish. If you can't find what you need, and it's very important to you, consider purchasing stock fasteners, then taking them into a plating shop for the finish. Look in the Yellow Pages under *Metals*, *Metal Finishing*, and/or *Metal Fabricators* (for starters) for local businesses near you.

You could, for example, purchase some basic 6-32 × 1" aluminum machine screws and take them in for anodizing in a color of your choice. You will be charged a base price, plus some cost based on either weight, surface area, material use, or a combination of the three.

Saving Money Buying Fasteners

When shopping for fasteners, you can save a considerable amount of money by purchasing in large quantities. At the hardware store a package of 10 #8 machine screws may cost 99 cents, yet a package of 100 may be priced at $3.99. The price difference is the cost of packaging smaller quantities.

If you think you'll make heavy use of a certain size of fastener in your robots, invest in the bigger box and pocket the savings. Of course, buy in bulk only when it's warranted, or you may end up with a garage full of fasteners you'll never use. A good approach is to first purchase only what you need in any new size. As you build your robots, you'll discover which sizes you use the most. These should be the ones you purchase in bulk.

Stocking Up: Selecting Common Fastener Sizes

Robot builders gravitate toward favorite materials, and fasteners are no exception. I can't tell you which sizes of fasteners to buy, because your design choices may be different from mine. But I can tell you what is used the most in my robot workshop. Perhaps that'll give you a starting point if you're starting to stock up.

For small tabletop robots I try to use 4-40 screws and nuts whenever possible, because they're about half the weight of 6-32 screws, and of course they're smaller. I use 4-40 × 1/2" screws and nuts to mount servos on brackets, and 4-40 × 3/4" screws for mounting small motors. Larger motors (up to about 6 to 8 pounds) can be fastened using 6-32 or 8-32 hardware.

For the small robots I try to keep the following fasteners in stock at all times:

- 6-32 nylon machine screws, in lengths 1/2" and 1"
- 8-32 × 1/2" and 1" nylon machine screws
- 4-40 steel machine screws in lengths 1/2", 3/4", and 1"
- 6-32 steel machine screws in lengths 1/2", 1", 1 1/2", 2", and 3" (I often use the 2" and 3" lengths to create risers for robots with multiple decks.)
- 8-32 steel machine screws in lengths 1/2" and 1"
- 6-32 × 3/4" wood screws, when fastening together panels of rigid expanded PVC
- 6-32 × 1/4" blind nuts
- 8-32 × 1/4" blind nuts

For all sizes in nylon and steel, I keep a corresponding stock of nuts, flat washers, and lock washers.

Note that nylon fasteners are not stocked at all hardware and home improvement stores, and selection may be limited for the size 4-40 hardware. Mail-order or hobby stores are two alternative sources for specialty fasteners.

Some tasks call for even smaller fasteners than 4-40. I keep on hand a small assortment of 2-56 hardware, typically 2-56 × 1/2" and 3/4" screws and nuts. Hobby stores that cater to the RC enthusiast are reliable sources for 2-56 fasteners.

For larger rover robots I keep a stock of the following:

- 10-24 (not 10-32) steel machine screws in lengths 1/2", 1", and 1 1/2"
- 1/4"-20 steel machine screws in lengths 1/2", 1", and 1 1/2"

Larger sizes are purchased when needed. Specialty designs, such as combat robots, require their own unique size assortments. For example, you may need a set of 7/16" × 2" stainless steel (for strength) bolts and nuts to fasten the body to the frame. These should be purchased on an as-needed basis.

Alternative Fasteners

Additional types of fasteners are at your disposal, should you need their unique properties. They include the following:

- *Pop or blind rivets* (see Figure 6-18) let you permanently assemble pieces. Use aluminum rivets for low weights. You need a pop rivet tool to use the rivets. Pop rivets are also called blind rivets because you do not need access to the back side of the rivet to secure them. The downside to pop rivets is that construction is permanent. You cannot disassemble the robot unless you drill out the rivets.

- *Arrow fasteners* are one-piece plastic designed to join two materials together. Arrow fasteners are constructed of nylon and are very durable. Many are designed to be only semipermanent: Using a screwdriver as a miniature crowbar, you can pop the fastener off, if needed. A disadvantage to arrow fasteners is that you must choose exactly the correct size for the thicknesses of material you are using.

In addition, refer to Chapter 5, "Mechanical Construction Techniques," for using hook-and-loop fasteners (such as Velcro and Dual Lock), plastic ties, double-sided tape, and adhesive transfer tape.

Brackets

Brackets are used to hold two or more pieces together, usually (but not always) at right angles. They come in a variety of shapes and materials.

Zinc-Plated Steel Brackets

Though intended to lash two pieces of wood together, hardware brackets are ideal for general robotics construction. These brackets are available in a variety of sizes and styles. You can use the brackets to build the frame of a robot constructed with various stock materials.

Figure 6-18 Pop rivets provide a permanent fastening solution and are easily applied using a riveting tool.

The most common brackets are made of 14 to 18 gauge steel. (The lower the number, the thicker the metal.) In order to resist corrosion and rust, the steel is zinc plated, giving the brackets their common metallic look. (Some brackets are plated with brass and are intended for decorative uses. They're more expensive and they have limited use in robotics.) Common sizes and types of steel brackets are

- 1½" × ⅜" flat corner brackets—used when joining pieces cut at 45-degree miters to make a frame.
- 1" × ⅜" or 1" × ½" corner angle brackets—used when attaching the stock to base plates and when securing various components (such as motors) to the robot.
- Same as above, but in sizes to about 2" × 2¾".

Keep in mind that angle brackets are made of fairly heavy steel and therefore are heavy. If you use several of them, they can add considerably to the weight or your robot. If you must keep weight down, consider substituting angle brackets for other mounting techniques, including gluing, brazing (for metal), or screws fastened directly into the frame or base material of your robot.

Specialty Furniture Brackets

The better hardware stores stock a small assortment of replacement parts for shelves, cabinets, and furniture. Among these parts are several sizes and styles of brackets, made of either metal or plastic. In the local True Value hardware

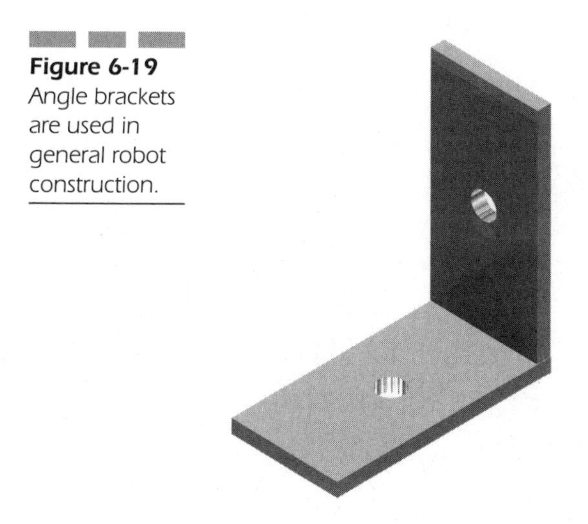

Figure 6-19 Angle brackets are used in general robot construction.

store near me, these parts are socked away in little yellow drawers. You buy them in single quantities, and prices are a little high.

Another source of handy brackets for robotics is the local cabinet-making shop. Though reselling parts isn't their main business, you never know what you can get unless you ask. Look for heavy-duty *knock-down* (KD) brackets, like that in Figure 6-20.These are used to lash together two pieces of heavy wood. With the right fastener, they'll hold together even heavy robots.

Knock down furniture is also sold by catalog retailers, such as Ikea. Check for availability of spare parts. Depending on the retailer, spare parts are only available as replacements for specific products, so you must know what to order before you can order it. One way around this dilemma is to find an article of furniture in the store that has the part you need. You can then ask if spare parts are available for it.

Plastic Gusset Brackets

Metal brackets can add substantial weight to a robot. Plastic brackets add little weight, but unless you're careful, they don't provide much holding power. Such is not the case with gusseted brackets, like those shown in Figure 6-21.

The bracket is made of plastic, such as *high-density polyethylene* (HDPE). This makes it very light. To add strength, the bracket uses molded-in gussets that reinforce the plastic at its critical stress points. The result is a bracket that is about as strong as a steel bracket but only a fraction of its weight.

Alas, plastic gusset brackets are not easy to find. They are available from some furniture-building outlets, as well as from select online resources, such as Budget Robotics. Sizes are fairly limited, but those sizes tend to be quite adequate for most jobs.

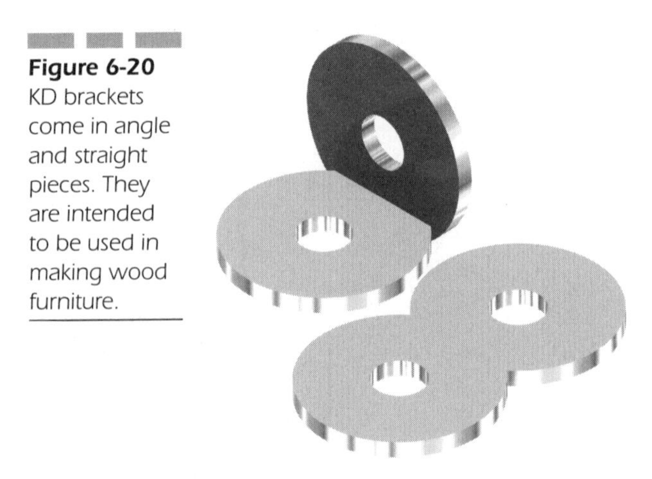

Figure 6-20 KD brackets come in angle and straight pieces. They are intended to be used in making wood furniture.

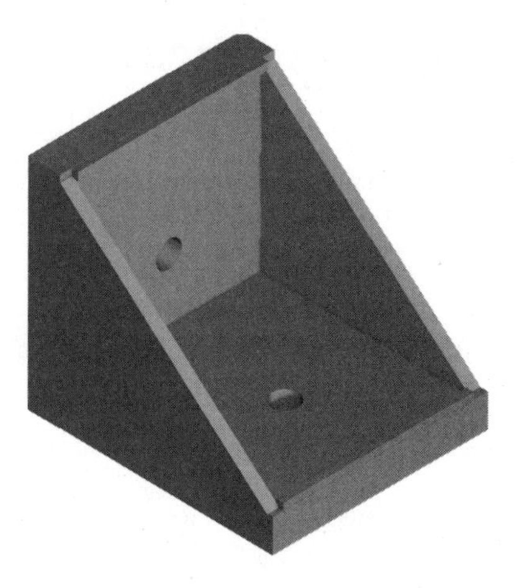

Figure 6-21 Plastic brackets with molded-in gussets are very strong yet lightweight.

Other Brackets

There are even more sources for brackets. They include the following:

- *Mirror clips* (hardware store) can be used as small brackets. Most mirror clips have only one hole and may not be in the familiar L-shape. You can always drill more holes and bend the clip to the shape you want.
- *Small metal L-brackets* (computer supply store) are used to construct electronics and computer systems. You'll have better luck finding these online.
- *Extra parts* of metal and plastic from Erector Sets and similar construction toys make for inexpensive brackets.
- *Self-made brackets* can be constructed from aluminum and brass metal, available at hobby stores and some hardware stores. Drill the holes, cut to length, and bend as needed.

Metal Stock at the Corner Hardware Store

Have a hankering to construct the next Terminator robot? You'll need about $100,000 in titanium to do it. A better idea: Make another robot and settle for commonly available metal stock, such as aluminum plate, channel, and rod. Your local hardware or home improvement store is the best place to begin. Here's what you'll find at the better-stocked stores.

Extruded Aluminum and Steel

Yet another form of useful metal is the extrusion, so-called because the metal is produced by extruding the heated, molten material out of a shaped orifice. The metal cools in the shape of the orifice, which can be a round tube; a square tube; and various L, T, I, and U shapes (to mention only a few letters of the alphabet). Metal extrusions can be used to construct the frame of a robot, for example, or to custom-make brackets and other parts. Refer to Figure 6-22 for some common extrusion shapes.

Almost all metals are available in extruded shapes, but for the purpose of robotics, the three most useful extruded metals are

- *Aluminum*, usually 6061 alloy, anodized with a brushed silver appearance, will not rust, but corrosion is possible if left outdoors. This material is reasonably easy to cut and drill, and is affordable. It's available at most hardware stores in various sizes and lengths. A typical aluminum extrusion is 1" equal-L angle; this means an L shape with 1" on each side. The typical thickness of aluminum extrusions is 1/32" to 1/8". Lengths are typically from 1 to 8 feet.
- *Mild steel*, for general house and yard work, is weldable, is uncoated, and will rust if not painted. This material is suitable for larger, heavy-duty robots, such as those meant for combat. The typical thickness ranges from 1/8" to 1/4".
- *Brass* and *copper* extrusions are available at some hardware and hobby stores in limited sizes and varieties. Most are round and square tubing, with thicknesses from 1/64" to no more than 1/16". You would use these when you need lightweight and solderable (or brazable) metal, such as for

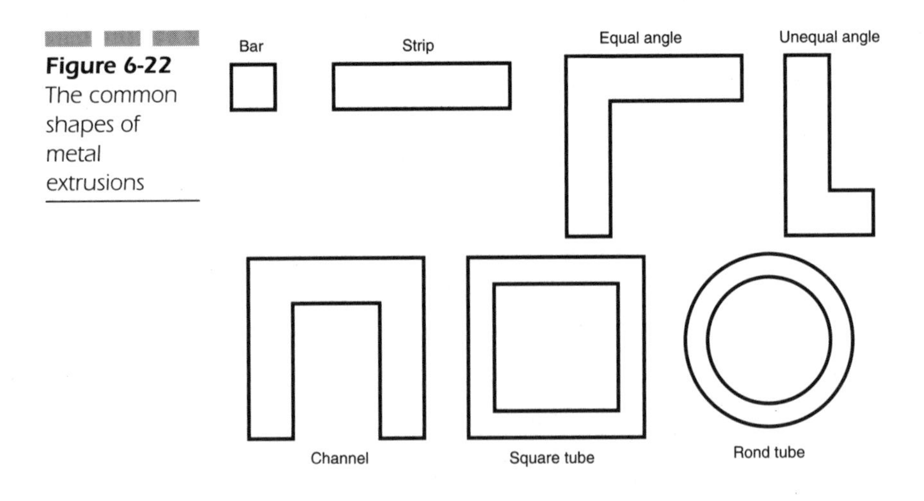

Figure 6-22 The common shapes of metal extrusions

constructing the legs of a small hexapod walking robot. Neither brass nor copper will rust if not painted, but both can tarnish.

Extruded aluminum and steel come in 2, 3, 4, 6, or 8 foot sections; some stores will let you buy cut pieces. Aluminum is lighter and easier to work with, but steel is stronger. Use steel when you need the strength, but otherwise, opt for aluminum.

Extruded aluminum and steel are available in more than two dozen common styles, from thin bars to pipes to square posts. Although you can use any of it as you see fit, a couple of standard sizes may prove to be particularly beneficial in your robot building endeavors:

- 1" × 1" × 1/16" angle
- 57/64" × 9/16" × 1/16" channel
- 41/64" × 1/2" × 1/16" channel
- Bar stock, widths from 1" to 3" and thicknesses from 1/16" to 1/4"

Mending Plates

The typical wood-frame home uses galvanized mending plates, joist hangars, and other metal pieces to join lumber together. Many of these metal pieces are weird shapes, but flat plates are available in a number of widths and lengths. You can use the plates as is or cut to size (the material is galvanized steel and is hard to cut; be sure to use a hacksaw with a fresh blade). The plates have numerous predrilled holes in them to facilitate hammering with nails, but you can drill new holes where you need them.

Mending plates (sometimes referred to as *Simpson ties*, after the name of a major manufacturer) are available in lengths of about 4", 6", and 12", by 4" or 6" inches wide, and also in 2" wide T shapes. You can usually find mending plates, angles, and other steel framing hardware in the nail and fastener section of the home improvement store.

Iron Angle Brackets

You need a way to connect all the metal pieces together. The easiest way is to use galvanized iron brackets, located in the hardware section of the store. Angle brackets come in a variety of sizes and shapes, and they have predrilled holes to facilitate construction. The 3/8"-wide brackets fit easily into the two sizes of channel stock previously mentioned. You need only to drill a corresponding hole in the channel stock and attach the pieces together with nuts and bolts. The result is a very sturdy and clean-looking frame. You'll find the flat corner-angle iron, corner angle (L), and flat mending iron to be particularly useful.

Using Extrusions as Framing

Perhaps the most common application for metal extrusions is building sturdy robot frames. As noted, common shapes are the U- and L-channels, with the U-channel being my personal favorite. A handy size is approximately $\frac{1}{2}'' \times \frac{1}{2}'' \times \frac{5}{8}''$, large enough to accommodate fasteners, L-brackets, and other hardware, but not so large that the metal unnecessarily adds to the weight of the robot.

To construct a frame, the extrusion is cut to length and then joined either to itself or to something else by way of brackets. Metal or plastic fasteners can be used as needed. For a lightweight robot, use either nylon fasteners (if the frame is fairly small, say under 8"), or 4-40 steel machine screws. You can also substitute aluminum pop rivets for screws, but this naturally makes the frame construction permanent.

Weight can add up quickly when using brackets, so choose the smallest available bracket that's consistent with the overall load bearing on the frame. Extra large steel L-brackets are not necessary for a robot under about 12" to 15". Opt instead for the smaller $1'' \times 1'' \times \frac{1}{2}''$ steel brackets or plastic brackets.

Using Extrusions as Brackets

Ready-made steel angle brackets are convenient and cheap for the bulk of any robotics project. But sometimes you need a size or shape that's not available at the corner hardware store. Or the typical steel bracket is too heavy, and an aluminum version would be perfect—only they don't have many brackets in aluminum. The solution: Cut your own brackets out of metal extrusions.

Most brackets have one or more holes per leg, so start by drilling the holes you'll need to mount the bracket. Remove any burrs or flash around the drill holes. Once drilling is complete, mark off the metal for cutting. An electric chop saw makes this job easy, but a hack saw and miter block can also be used.

Figure 6-23 A robot frame can be made from hardware store aluminum extrusions.

Caution! By their nature brackets are small and, therefore, dangerous to cut on an electric saw without using some form of hold-down clamp. Clamp the extrusion on *both sides* of the blade. A small C-clamp or spring-loaded clamp should be sufficient. The idea is to use the clamp, and not your bare fingers, to hold the small bracket while it's being cut from the main piece. *Cut off the metal, not your fingers!*

Even More Hardware Store Finds

Take a stroll down the aisles of a well-stocked hardware store and you're sure to find plenty of parts you can use for your robot creations. It's impossible to list all of the goodies you'll likely find, and even less practical to discuss every conceivable use, but the following is a sample list to get your creative juices flowing:

- *Rubber grommets* protect wiring, but also act as springy material for whiskers, sensors, and linkages.
- *Cable clips* (Figure 6-24) for wire and small piping and tubing (e.g., aquarium tubing) are useful for wiring containment, as well as for attaching parts (small motors, sensors, etc.).
- *Large wire terminals* can be used for crimping cables for grippers and leg mechanisms in addition to being used for electrical applications.
- *Springs* have 1,001 uses, including in touch sensors, bumpers, and even robot decoration.
- *Rubber and metal feet* for small furniture and appliances are ideal for walking pads for legged robots.

Figure 6-24 Plastic cable clamps can be used for parts mounting.

- Use *metal conduit and electrical metallic tubing* (EMT) fittings in the electrical section to make a very heavy-duty frame for a larger 'bot.
- *Shower and patio door parts* include rollers (nylon or ball bearing) for use as small casters and wheels.
- *Weather stripping and rubber door insulation* is perfect for robot bumpers (Figure 6-25).

Radio Control (RC) Model Parts and Components

When you think about it, a small mobile robot is not much different from an RC model car, boat, or airplane—except a robot's activity is largely controlled by a computer rather than remotely by a human being. The mechanical aspects of the robot have many of the same parts and components as RC models do. Because RC modeling is such a popular hobby, practiced all over the world, a vast array of products are widely available, and at affordable prices.

Pushrods, Cables, and Linkages

Occasionally used in amateur robotics, but still very useful, is the pushrod, which is basically a piece of heavy, thick metal. More often than not, the *pushrod* is threaded on one end, so that it firmly connects into a *clevis*, *swivel ball*, or other linkage fitting. The other end of the pushrod is bent to make a hook and is attached to the servo by way of a servo horn (more information on servo horns will follow). See Figure 6-26 for an assortment of pushrod and clevis hardware.

You can use pushrods in robotic designs that require you to transfer linear motion from one point to another, such as in an arm or finger grippers. This allows you to place the heavy and bulky servo in the base of the arm.

You must match the sizes of the pushrod and the clevis. You can choose from plastic or metal clevises; the metal variety can be either screw-on or solderable.

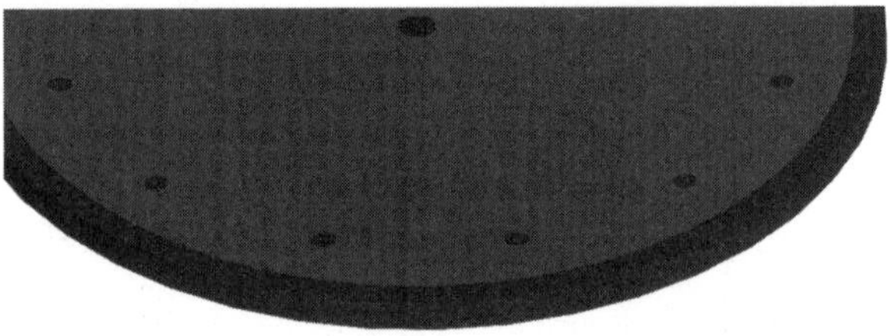

Figure 6-25 Use weather stripping to make robot bumpers and cushions.

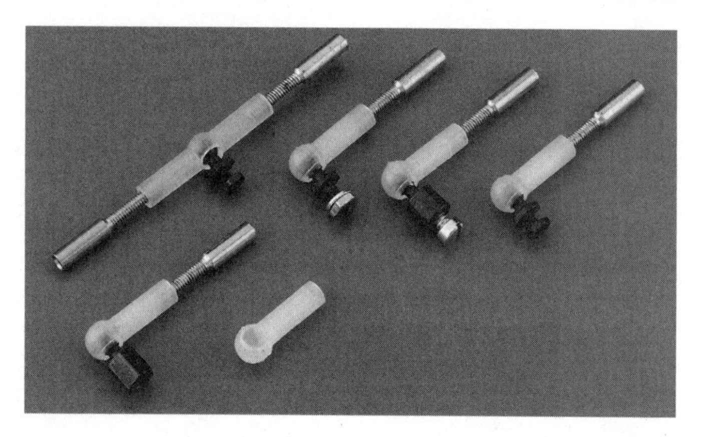

Figure 6-26 Pushrods and clevis linkages find many uses in robotics.

The screw-on type is definitely easier to work with, but requires a threaded pushrod, and this adds to the expense.

Flexible cables, with and without outer plastic sheathings, are used in much the same way as pushrods, but are useful when the linkage cannot be rigid.

Servo Horns and Bell Cranks

Servo horns fit onto the shafts of RC servos and are most often used to convert the rotational movement of the servo to a linear movement. In a model airplane, for example, this linear movement might be to move the ailerons up and down. Servo horns come in a variety of shapes and sizes; it's best to simply take a look at what's available and choose the kind you think will work best. You need to get servo horns for your make and model of servo, because the mounting hole varies in size.

By the way, there's also plain ol' *control horns*, which are not for attaching to the output shafts of servos. Control horns are most often used on the other end of the mechanical linkage from the servo—you would use a control horn on the surface of an aileron.

Bell cranks are similar to control horns, and they serve as levers. They are often used to enlarge or reduce the amount of linear movement.

Dura Collars

Though used for many applications, dura collars are typically employed to keep RC wheels on their axle shafts. The collar is made of plated metal and is drilled for a set screw. You use a hex wrench to tighten the collar around the axle shaft. Dura collars come in a variety of sizes; match the collar to the diameter of the axle shaft. Common sizes are 1/16" (1.5 mm) to 3/16" (4.7 mm).

Miscellaneous RC Airplane Hardware

It's not possible to describe every piece of hardware available for RC modeling, but the following is a quick rundown of some of the more useful components:

- *Control hinges* are plastic or metal hinges with mounting holes. They come in a variety of sizes. For robotics, go for the best you can get so the hinge doesn't fall apart from extra wear.
- *Bolt, nut, and screw hardware* is available in a variety of sizes, from tiny 2-56 threads to standard 6-32 and 8-32 threads. Look for blind nuts (Figure 6-27), which let you mount things like servos and motors flush to the body of the robot. Blind nuts need to be used with soft plastic, wood, or similar materials (but not metal or hard plastic like acrylics). The fins of the nut must be able to dig into the material to provide a secure fit.
- *Threaded inserts* are also used with softer materials and screw into the material to create a standard-size threaded hole.
- *Servo tape* is a wide and supersticky tape for holding servos, batteries, and other objects to a frame of a robot.
- *Hex socket head screws* are precision machined and have threads from 2-56 through 6-32. They have hex-socket heads, sometimes with a knurled knob to allow for easier manual tightening.
- *Threaded couplers* extend the length of threaded rods by allowing them to be connected together.
- *Replacement servo gears,* though intended to repair broken or worn-out servos, can also be used for any other purpose you choose. Replacement gears may be plastic or metal, depending on the make and model of servo they are for.

Figure 6-27 Blind nuts bite into the material.

Specialty Parts for RC Racing

RC racing—cars, trucks, even boats—offer yet more options for small parts. Prices are quite varied, from dollar finds to specialty custom-made parts that cost a small fortune. Shop carefully!

RC vehicles are sold in various scales, starting at 1/10 for fairly large models (about 12" to 18" long), and 1/48 and smaller, for miniature cars and trucks. Because of their more rugged design, the off-road cars and buggies tend to offer the better assortment of usable parts. Among the most useful RC racing parts are the following:

Shocks and shock kits	These are spring-loaded and look like automobile shocks. Use them to create bumpers for robots or to construct an elaborate suspension system.
Bearings and bushings	Less expensive than industrial versions of the same things, use model racer bearings and bushings for robot drive trains. Example prices: about $6 per ball bearing set, as opposed to $12 to $15 for similar size bearings from an industrial supply catalog.
Chassis bumpers	Available in numerous sizes and shapes, many can be adapted as bumpers for a robot. The choice of materials includes metal (typically aluminum), unbreakable plastic, and carbon composite. Some bumpers are the perfect size and shape as a scoop for a sumo robot. Skid plates are also useful for this application.
Transmission components	Model cars, trucks, and buggies use a variety of novel power-transmission drive components you can reengineer for your robot. Examples include *constant-velocity drive* (CVD) kits, drive axles, one- and two-speed transmissions, universal joints, steering bell cranks and mechanisms, belt and pulley drives, gear sets, and more. You'd have to be an expert machinist to replicate these parts, and similar items from an industrial supply outlet would be considerably more expensive.
Wheels and hubs	Wheels for RC cars and trucks are ideal for small robots. They're the right diameter and you can select from a wide variety of tire materials—foam, rubber, urethane, and others. The downside is attaching the wheels to your robot's motor. When using a dc gear motor, your best bet is standard wheel hubs that use a set screw. You may need to drill out the hub to match the diameter of the motor drive shaft. For RC servo motors, one viable technique is to attach the wheel using a large servo disc and fasteners.
Electrical power components	Speed controls (especially those with forward and reverse), servos, and batteries for RC vehicles are inherently ideal for use in robotics.

Parts for Model Railroading

Some model railroad components are also handy in the robot building arts. Most hobby stores segregate the model airplane parts from the model railroad parts. If you don't see what you're looking for in one section of the store, try another. Because model railroads come in various scales and gauges (basically different sizes), the accessories and parts likewise match. The most common scales and gauges, in descending size, are G, O, S, HO, and N. Not all stores carry all varieties.

In the model railroad section, be on the lookout for

- *Backdrop construction materials*, including lightweight plastics, rubber and plastic castings, and spray foam.
- *Drive motors*, often in unusual shapes and arrangements, for locomotives. These are very high quality with great efficiency in a miniature scale. Many come with gearboxes attached or can be readily attached to a gearbox.
- *Mechanicals*, such as trellises, crossing arms, and so forth. Some are motorized and can be adapted to robot use.
- *Rail joiners* and *track,* which are made of a semisoft metal that can be bent to shape.
- *Molding and casting supplies*, such as plaster cloth and latex rubber.
- *Miniature electronics*, such as three-way signal lights, which can be adapted to some robotic designs.
- *Armatures*, which are originally intended to create trees and shrubbery. They consist of semistiff bendable wire. Use armature materials to form shapes, or create bumpers and whisker sensors.

CHAPTER 7

Bases and Frames

When you look at that classic 1957 Chevy Bel Air, what do you see? Odds are it's the bold front grille, the haughty brow of the headlamps, the curvaceous windshield, or the playful rocket engine fins. That's the stuff Detroit wanted you to see, after all. But under the flashy chrome and brightly painted tin is the lowly —nah, boring—frame that holds it all together. Just about everything important attaches to this frame: the motor, the chassis, the transmission. Without this frame, the Bel Air would be a pile of junk.

No matter what design you choose for your robot, its structure is the key to its success. Fortunately, unlike a car frame, your robot's frame needn't be cross-welded and powder coated to resist rust. It can be as simple as a piece of plastic or a length of aluminum tubing glued together with epoxy.

In this chapter, we discuss the core construction of workable and near-universal bases for wheeled and tracked robots. Topics covered include using wood, plastic, and metal materials, and how to best work with these to fashion quality parts for your robots. Metal and plastic extrusions play an important role in building sturdy robot frames, and these are covered in some detail. Also, this chapter presents the concept of building metal robots without cutting to size.

A moment now to review terminology: I use the words *base* and *frame* to denote the main structural body of the robot. In most cases, I treat them as distinct concepts: The base being a combination structure and mounting platform, while the frame is an undercarriage upon which additional body parts may be attached.

However, for the sake of material selection and workshop procedures, there is no difference at all between a base and a frame. Some materials, such as metal bars, are obviously best used to build frames, while others—say, rigid plastic sheets—are functionally superior as bases.

In Review: Selecting the Right Material

Before construction begins, you must select the principal construction material you wish to use for the base of your robot. For this book I've limited the materials to various types of wood, plastic, and metal. These materials, and the many choices within each category, are more fully discussed in previous chapters; see Table 7-1 for a quick recap of the advantages and disadvantages of each.

There is no single ideal material for constructing robots. Each project requires a review of the following:

- **The robot itself** Determine especially its physical attributes (e.g., large, small, heavy, light).

Table 7-1 Review of construction materials

Material	Pros	Cons
Wood	Universally available; reasonably low cost; easy to work with using ordinary shop tools; hardwood plywood (the recommended wood for most robot bases) very sturdy and strong	Not as strong as plastic or metal; can warp with moisture (should be painted or sealed); cracks and splinters under stress
Plastic	Strong and durable; comes in many forms, including sheets and extruded shapes; several common types of sheet plastic (acrylic and polycarbonate) readily available at hardware and home improvement stores; other types can be purchased via mail order at reasonable cost	Melts or sags at higher temperatures; some types of plastic (e.g., acrylic) can crack or splinter with impact; PVCs and styrenes not dimensionally stable under stress; exotic types hard to find; better plastics are expensive; some specialty tools required for professional-looking cuts and holes
Metal	Very strong; aluminum available in a variety of convenient shapes (sheet, extruded shapes); dimensionally stable even at higher loads and heats	Heaviest of all materials; requires power tools and sharp saws and bits for proper construction; harder to work with (requires more skill); can be expensive

- **The tasks the robot is expected to do** Robots that do not perform heavy work (lifting objects or smashing into other robots) do not need heavy-duty materials.
- **Your budget** Everyone has a limit on what he or she can spend on robot materials. Tight budgets call for the least expensive materials.
- **Your construction skills** Wood and plastic robots are easier to build than metal ones.
- **Your tools** Building metal robots requires heavier-duty tools than when building wood or plastic 'bots.

Wood Bases and Framing

Wood is among the oldest construction materials on the earth, and arguably the most useful. It even has applications in robot building. Wood is fairly cheap and easy to find, and it's easy to work with using common tools.

Wood Robot Bases

Bases made of wood are best constructed using hardwood plywood. These are available in various thicknesses, with 1/4" being a good all-around choice for a robot under 10" in diameter. Even smaller robots can use the 1/8" or 5/32" thicknesses. Hardwood plywood is routinely available at craft and hobby stores, as well as at many home improvement stores. You can buy it in convenient cut-down sheets of 24" × 24" (and even smaller).

For a square base, you need only cut the wood to size using a hand or power saw. A power saw with a guide fence (e.g., a table saw) is preferable, as this will yield the straightest cut. You can mark the desired dimensions right on the surface of the wood using a #2 pencil.

NOTE: *Don't assume the plywood pieces you buy are cut square. Verify and trim the sides to ensure squareness.*

For a shaped base—round, oval, or something else—a scroll saw or band saw will make short work of cutting the wood. Draw the desired shape on the wood using a pencil, and then follow the lines during cutting. With some practice, you can even cut round bases, though a better approach is to use a small router or hobby tool and a circle-cutting attachment. Drill a hole for the center of the base, and screw the attachment in place. Position the router at the desired distance from the center, and make the cut.

Wood Robot Frames

Wood frames can be constructed using strips of hardwood. You can buy the strips premade, or cut your own if you have a table saw. For frames less than 12" square, strips 3/8" to 5/8" wide are adequate; use 3/4" or even 1" wide wood for larger robots.

Wood selection is critical. Stay away from softwoods, such as pine, fir, and redwood. They are not strong enough except for in the smallest of bases. Several good all-around hardwoods that are available throughout most of North America are alder, ash, American beech, and poplar. These woods are also among the least expensive.

Square frames can be constructed using miter cuts; nail, staple, or glue the corners together, or better yet, use flat corner brackets (see Figure 7-1) for a rock-solid construction. The brackets can be secured using wood screws or (preferred) machine screws and nuts.

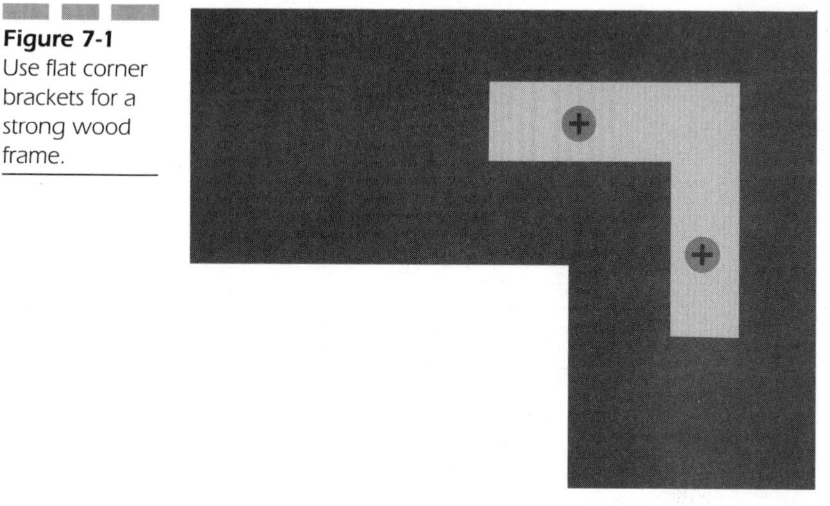
Figure 7-1 Use flat corner brackets for a strong wood frame.

Box frames can be constructed using two (or more) square frames, anchored together with plywood or plastic panels, as shown in Figure 7-2. As needed, cut out segments of the panel to save weight. Avoid removing too much material or the panel will be weakened. The panel can be affixed to the frame using wood screws or machine screws and nuts.

Wood Finishing

You can extend the life of your wood robot bases and frames, not to mention enhance their looks, with simple finishing. Wood finishing involves sanding, which smooths down the exposed grain, and then painting or sealing. Small pieces can be sanded by hand, but larger bases benefit from a power sander.

Sandpapers are available in a variety of grits—the lower the grit number, the more coarse the paper. The recommended approach is to start first with a coarse grit to remove splinters and other rough spots, and then finish off with a moderate- or fine-grit paper. For wood, you can select between aluminum oxide and garnet grits. Aluminum oxide lasts a bit longer. Sandpapers for wood are used dry. For hand sanding, wrap the paper around a block to provide even pressure over the wood. See Table 7-2.

Wood can be painted with a brush or spray. Brush painting with acrylic (available at craft stores) produces excellent results with little or no waste. One coat may be sufficient, but two or three coats may be necessary. Woods with an open grain may need to be sealed first using a varnish or sealer, or else the paint will soak into the wood, no matter how many coats you apply. You may also opt to skip the painting step altogether and apply only the sealant.

Figure 7-2 To create a box frame, use two square frames connected with body panels.

Table 7-2 Recommended sandpaper grits for wood finishing

	Grit			
Use	**F**	**M**	**C**	**EC**
Heavy sanding			X	X
Moderate sanding		X		
Finish sanding	X			

Grit Key	Name	Grit
EC	Extra coarse	30-40
C	Coarse	50-60
M	Medium	80-100
F	Fine	120-150

Plastic Bases and Framing

My unabashed favorite material for building robots is plastic. For its size, plastic is stronger than most woods, easier to work with, and cheaper than metal. Preferred types of plastics include expanded *polyvinyl chloride* (PVC) sheets, ABS, polycarbonate, and acrylic. Expanded PVC is used in industry as an alternative to wood (e.g., wood molding), and it cuts and drills much like wood.

Plastic Robot Bases

With the exception of acrylic and polycarbonate sheets, the plastics mentioned must be purchased from specialty plastics retailers or ordered through the mail from suppliers such as McMaster-Carr. Sheet segments of 12" × 12" and larger are routinely available. If you have access to a local plastics retailer, you can even buy the plastic in the full 48" × 96" sheets.

When possible, use a power saw to cut plastics. The ideal blade for cutting acrylic and most other plastics is specifically designed for the job and is labeled as such. The use of a table saw is highly recommended. When a specialty acrylics blade is not available, use a hollow ground high-speed blade with no set (the teeth do not alternate to the left and right) and with at least five *teeth per inch* (tpi). Carbide-tipped blades with a triple chip tooth will give the smoothest cuts. When using a table saw, set the blade height about 1 /8" above the height of the material. This will reduce edge chipping. Polycarbonate, ABS, and PVC plastic is cut with a similar blade.

Using any kind of reciprocal saw (jig saws, scroll saw, etc.) is not recommended for plastics because the blade can overheat the material, causing it to remelt into the cut.

For more elaborate cuts and designs, use a router or high-speed hobby tool outfitted with a milling bit. The bit should be 2- or 4-flute, with a 30-degree helix. A standard end-mill (see Figure 7-3) is adequate. For best results when working with harder plastics (e.g., polycarbonate and acrylic), the bit should be no less than half the diameter of the material. That is, if the material is 1/4" thick, the bit should not be less than 1/8".

When routing plastic it's important to eliminate all vibrations. Be sure the material is held down firmly—but not with your hands! Otherwise, the plastic will vibrate against the bit, making for a very rough cut. As with wood, you can cut circular bases using a circle-cutting attachment. Work slowly to avoid overheating the bit.

For thin acrylic stock (1/16" and smaller), you can make straight cuts by scoring. Using a metal straightedge and sharp utility knife, make a score along the entire length of the cut. Place a 1/4" dowel directly under the score. Press down evenly on both sides of the score, and the plastic should snap apart. This method does not work as well with PVC and other plastics that are flexible.

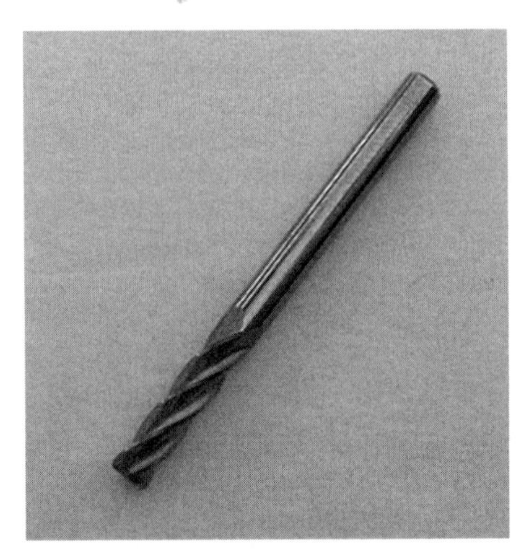

Figure 7-3 An end-mill suitable for plastics and many other materials. It can be used with a router or motorized hobby tool.

Plastic Robot Frames

The majority of plastics available to consumers are rather flexible. This makes them less than adequate for the job of creating robot frames. However, given a thick enough material—½" or larger—the flexing is minimized. You can cut the ends of the frame pieces to 45 degrees with a miter cutter, and assemble using flat angle brackets.

Another approach, if you have the benefit of a local plastics retailer, is to raid their surplus bin for lengths of acetal resin (also called Delrin). Acetal resin is very strong and sturdy, and resists flexing. Its typical use is in manufacturing parts, and it is provided in large slabs that are cut up for machining. Find lengths of 1" square material and cut them down to make frames.

Plastic Finishing

Most plastics do not need finishing. However, you may wish to smooth out any rough edges of cuts in the plastic by applying a light sanding with fine-grit aluminum oxide (not garnet) sandpaper, used wet or dry. If needed, you can also shape corners and edges of the plastic with a course or extra coarse sandpaper. See Table 7-3.

The edges of acrylic can be finished using a technique called flame burnishing, but this isn't recommended unless you know what you're doing. The process involves heating the plastic with the flame of a torch. Extreme care must be taken to keep the plastic from melting or even igniting! Instead of flame burnishing, use the sandpaper method described previously, or outfit your drill

Table 7-3 Recommended sandpaper grits for plastic finishing

Use	VF	F	C	EC
Shaping			X	X
Smoothing	X	X		

Grit Key	Name	Grit
EC	Extra coarse	30-40
C	Coarse	50-60
F	Fine	120-150
VF	Very Fine	160-200

motor with a cloth polishing wheel. Buy some polishing rouge from your local plastics outlet (or ask for some from a nearby jeweler) and burnish the edges with the rouge. It's much safer.

Likewise, most plastic does not need to be painted. However, this shouldn't stop you from painting your plastic robots, should you want a different look.

- Paint the top side of the plastic for a matte finish (clear acrylic or polycarbonate).
- Paint the bottom side of the plastic for a glossy finish (clear acrylic or polycarbonate).
- Spray painting yields the most professional results. Avoid heavy coats, or else the paint will sag.
- Use a solvent-based paint (but always test a small portion of the plastic first!). Examples include Testor's model spray paint. The solvent penetrates into the plastic, making the paint more permanent.
- Use an airbrush for an artistic paint job.

Metal Bases and Frames

The main benefit of a base or frame made of metal is its longevity. Even if your robot falls off the table or gets mauled by the dog, at least its body will remain intact. Metal provides a resiliency that wood and plastic cannot match, but it's harder to work, costs more, and weighs more.

Metal Robot Bases

Sheet metal can be used to make robot bases. Mild steel and aluminum are the most common for use in robots. For both, select a thickness that will support the size and weight of the robot, but without adding undue weight. Consider the following starting points in your design.

For small robots, under 8":

- Aluminum: 1/32" to 1/16" (0.03125" to 0.0625")
- Mild steel: 22 to 20 gauge

For medium robots, 9" to 15":

- Aluminum: 1/16" to 1/8" (0.0625" to 0.125")
- Mild steel: 20 to 18 gauge

For large robots, 16" to 24":

- Aluminum: 1/8" to 1/4" (0.125" to 0.250")
- Mild steel: 18 to 16 gauge

Note that because steel is stronger than aluminum, it does not need to be as thick to provide the same support. Twenty gauge steel is approximately 0.035", or a little larger than 1/32".

For thicknesses under 1/8" (0.125") for aluminum, or 18 gauge for steel, you can use aviation snips (see Figure 7-4) or a nibbler tool to cut and shape the metal. When cutting out circular bases, consider whether you'll use a right- or left-hand tool; it'll make better cuts. Right-hand and left-hand, in this case, refer to the direction of the cut, not whether you are right or left handed. For example, a left-

Figure 7-4 Aviation or aircraft snips come in straight, left-hand, and right-hand configurations.

hand snip is used to cut the circle counterclockwise (to the left). Bear in mind that the cheaper tools cannot cut material that thick. Depending on the tool, you may be limited to cutting 22 gauge stock.

For thicker stocks, specialized metalworking tools are needed. A metal shear is used to make straight cuts. Very thick steel can be cut with a torch and then ground down as needed with a motorized grinder. Heavy-duty pneumatic snips can be used with heavier stocks—up to about 16 gauge.

On the other side of the spectrum, the thinner aluminum stocks (1/32") can be cut to shape using a router equipped with an end-mill. A mill with two or four flutes and a 30-degree helix is appropriate. Secure the metal to the worktable so there is no vibration. A circle-cutting attachment can be used to cut out round bases.

Metal Robot Frames

Robot frames can be constructed using metal extruded shapes, available at most home improvement stores, and with specialty machine framing. Machine framing employs lengths of aluminum attached together with various connectors to produce frames of all shapes and sizes. We cover the subjects of metal extrusions and machine framing in depth later in this chapter.

Metal Finishing

After cutting, the metal should be ground to remove flash and burrs. Use a flat file to smooth out any rough edges. If there is a lot of material to remove, use a small grinding wheel attached to a drill motor or hobby tool.

Use sandpaper for a very smooth edge. This may be desirable for looks, but also for function. The metal pieces may need to slide against one another, and the surfaces need to be like glass. You can use aluminum oxide or emery cloth sandpapers for aluminum or mild steel. For general deburring and cleaning, use a fine or medium aluminum oxide paper; for finishing and polishing, use a fine emery cloth. See Table 7-4.

Table 7-4 Recommended sandpaper grits for metal finishing

Grit Key	Name	Grit
M	Medium	80-100
F	Fine	120-150

For a final smooth finish, buff the metal using 00 to 000 steel wool.

Bases and frames of mild steel should be painted to avoid rust. No painting or other treatment is needed if the aluminum you used is already anodized—it'll have a silvered, black, or colored satin appearance. For bare aluminum, the metal can be left as is, but you may prefer to paint it. There are several alternatives for painting aluminum and steel:

- *Brush painting* is simple and effective, but brush marks are hard to avoid, and it's easy to apply the paint too thick.
- *Spray painting* with a paint designed for metal yields better results, but only if you apply the paint in several light coats. Apply a primer coat first. Enamel paints work the best. Paints for automobile engines are especially durable. Let dry completely before handling.
- *Powder coating* is a process that applies a layer of paint to the metal using an electrostatic charge. The pigment penetrates the surface layer of the metal for a more durable finish. Powder coating can be performed at home if you have the right gear, but this is usually something for specialty paint shops.

In all cases, it is extremely important that the metal be completely clean and free of grease, oils (including skin oil), and dirt before painting. Use denatured alcohol to clean the metal, including all the nooks and crannies.

For bare aluminum, consider having the base or frame anodized. Look in the Yellow Pages under *Metal-Fabricators*, as well as under other *Metal* subcategories. You have a choice of several anodizing colors, including silver, black, blue, yellow, blue, and green. The anodizing process requires dipping the aluminum into vats of liquid. Therefore, you must remove all electronics and other parts from the base or frame, or they'll be ruined.

Aluminum Profiles

Machine framing is used to build the framework for machines used in commerce and industry. The material is extruded into unique *profiles*. The typical machine framing uses flanges that make it easy to secure other connectors and components. Figure 7-5 shows several representative cross-sections of machine framing. Several manufacturers offer aluminum extrusions, each one with slightly different shapes. Extrusions from one company will likely not work with that of another. It's best to purchase the aluminum profile and connectors together from the same source, so you can be sure of the fit.

Framing profiles come in a variety of sizes, and usually in two classes: *fractional* (also called *inch*) and *metric*. The choice depends on what measurement system you're most comfortable working in; besides small variations in size, there is no difference between the profiles. From here on out, we'll assume frac-

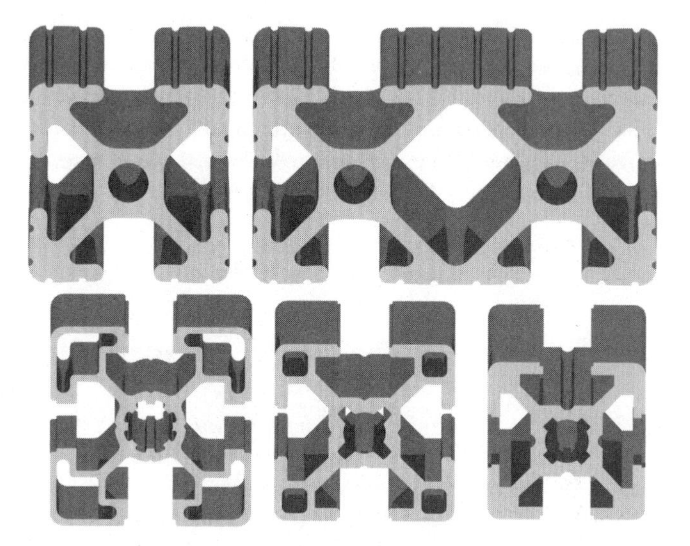

Figure 7-5 Some of the many styles of aluminum machine framing.

tional for our discussion, but note that similar products are available in equivalent metric sizes as well.

Machine framing components are first measured by their cross-section dimensions, such as 1" × 1", or 1½" × 2". For most robots under 18", the smaller material is sufficient, and it's less expensive all around. Sizes of up to 3" × 6" are available, when maximum structural strength is required.

The flanges of the machine-framing piece create what's commonly referred to as T-slots; when viewed end-on, the slot looks like a T. The T-slot provides a convenient method of attaching components to the framing. Special T-nuts, as well as various bolt-on clamps, which tighten against the slotted structure, can be used.

Beyond size, literally hundreds of profiles (i.e., extrusion shapes) are available, with the T-slot being the most common. In most profiles, a T-shaped slot is extruded on all four sides of a square tube. The benefit of the T-slot is that a component can be moved to any position along the length of the frame simply by loosening the fastener, sliding it up or down, and retightening it.

Also available are hexagon and octagon machine profiles. These are used when more elaborate constructions are required, because they allow for greater flexibility in attaching components to the T-slots. Rather than components connecting at right angles only, hexagon and octagon profiles allow for attaching components at 45- and 60-degree angles.

The framing pieces are cut to length with a hacksaw or power miter saw, and then assembled with connectors and other hardware. The typical machine framing construction does not require drilling, tapping, or welding, yet the finished product is quite strong. This is the main attraction of the system. Everything can be fastened tight together using the supplied connectors and hardware. What's more, the frame can be readily disassembled if you need to change your design.

Figure 7-6 shows a simple rectangular frame constructed of aluminum profile extrusion and plastic corner connectors. Construction time, including cutting and deburring the aluminum, is about 10 minutes. The connectors are removable, but in most cases, the friction fit between them and the metal is strong enough to hold the frame together. However, you can drill small holes through the metal and connector, and then secure the corner pieces with self-tapping screws. I recommend using sheet-metal screws, rather than wood screws, as the latter does not have threads in the shank portion of the screw.

Managing the Costs of Aluminum Profile

Although aluminum profile extrusions offer a fast and efficient way to build the frames and other mechanical systems of medium and large robots, they can be expensive. Most extrusions are sold by the foot or the inch. A 6-foot length of 1" by 1" aluminum T-slot costs an average of about $15 to $20. A similar length of extruded aluminum channel stock, available at the home improvement store, is barely half that amount. Of course, to use it you must drill mounting holes as needed.

Costs can be controlled by carefully selecting the pieces you need, in the exact lengths you need, so there is no waste. A foot of wasted extrusion can cost several dollars, so it's wise to make your investment stretch as far as it can. Although

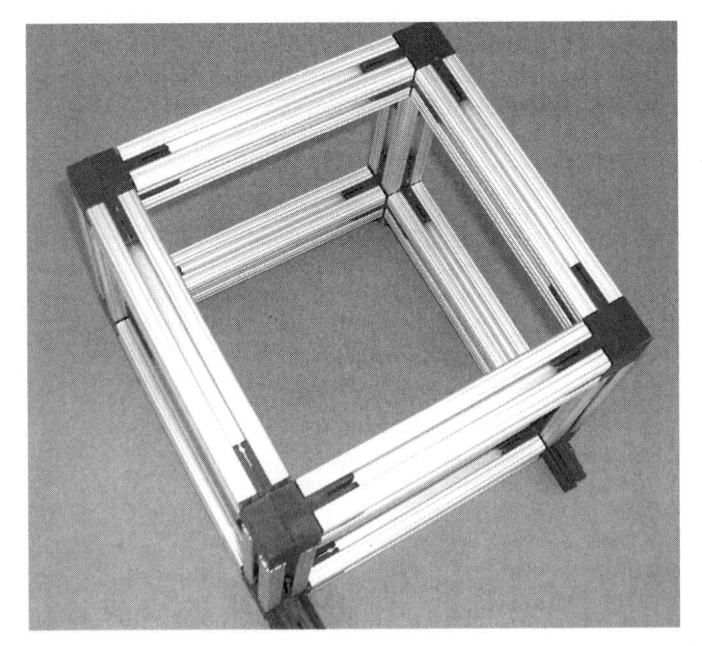

Figure 7-6 A basic robot frame constructed using machine-framing components.

this seems obvious, the task is made a bit more difficult because most extrusion is available in 6-foot, 8-foot, or 10-foot lengths (with 8-foot and 10-foot lengths most common). If you need only 4-foot, at best you'll have 2-foot extra (from a 6-foot length), for which you spent money unnecessarily.

Several sources of machine framing components sell precut pieces. This not only lets you get lengths more appropriate for your use, but also saves on shipping; it can cost more to ship a 6-foot or 8-foot length of extrusion than the price of the material. You may pay $15 for a 6-foot length, but then pay $20 to $25 to ship it (even surface shipping). Buying precut pieces also saves you from cutting and deburring them yourself, and the cuts are usually more accurate than you can make with a hacksaw.

On the downside, precut pieces cost more, so shop carefully. Weigh the price of buying the material in standard lengths (and paying more for shipping and for excess) against paying more for precut lengths.

Assembly with Connectors, Hinges, and Gussets

Basic frames are constructed using connectors, fasteners, or hinges. A square frame can be made using four lengths of profile extrusion and four *two-way L-connectors* (see Figure 7-7). A box frame can be constructed using 12 lengths of profile extrusion and eight *three-way Y-connectors*. You can vary the dimensions of the box by changing the lengths of the side, end, and riser pieces. For example, to make a 12" × 12" × 12" frame, cut each of the 12 extrusions to about 10". See Figure 7-8.

Other useful hardware for machine framing includes:

- *Hinges* offer added flexibility by allowing you to set the angle of the joined pieces. The hinges can be tightened to stay in place.
- *Gusset brackets* provide extra strength for perpendicular joints. The brackets are reinforced for extra strength.

Selected Sellers and Manufacturers of Machine Framing

The following are several manufacturers and suppliers of machine framing components:

80/20 Inc. ***www.8020.net***

Bosch Automation Products ***www.boschautomation.com***

Flexible Industrial Systems, Inc. ***www.goflexible.com***

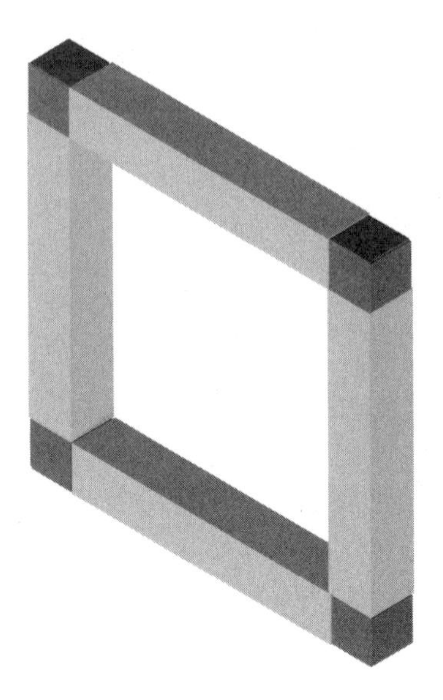

Figure 7-7
A square frame is made with four lengths of extrusion and four connectors.

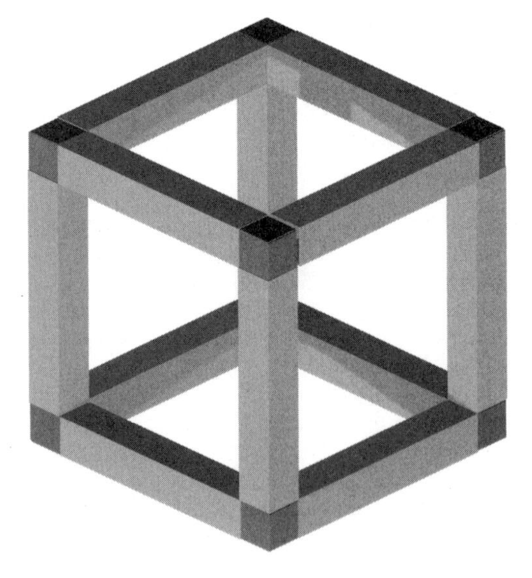

Figure 7-8
A box frame is made with 12 lengths of extrusion and eight connectors.

Frame World/Barrington Automation ***www.frame-world.com***

Kee Industrial Products, Inc. ***www.keeklamp.com***

Techmaster, Inc. ***www.techmasterinc.com***

Textube Corporation/Creform ***www.textube.com***

Alternatives to Extruded Machine Framing

If machine framing in plastic or aluminum is too rich for your blood, but you like the idea of assembling framing using tubing and connectors, consider display-fixture hardware designed for retail stores. Available in plastic, aluminum, or steel, this round or square material can be cut to length, and then joined using various connectors.

One common style is 1½" PVC tubing, available in white, gray, black, and other colors. The tubing connects with various shapes of corner connectors (Figure 7-9). The following are a few examples:

- A two-way L joins two pieces of pipe at a right angle.
- A three-way T joins three pieces of pipe: two pipes connected end to end, and a third is connected as a riser.
- A four-way T joins four pieces of pipe in an intersection arrangement.

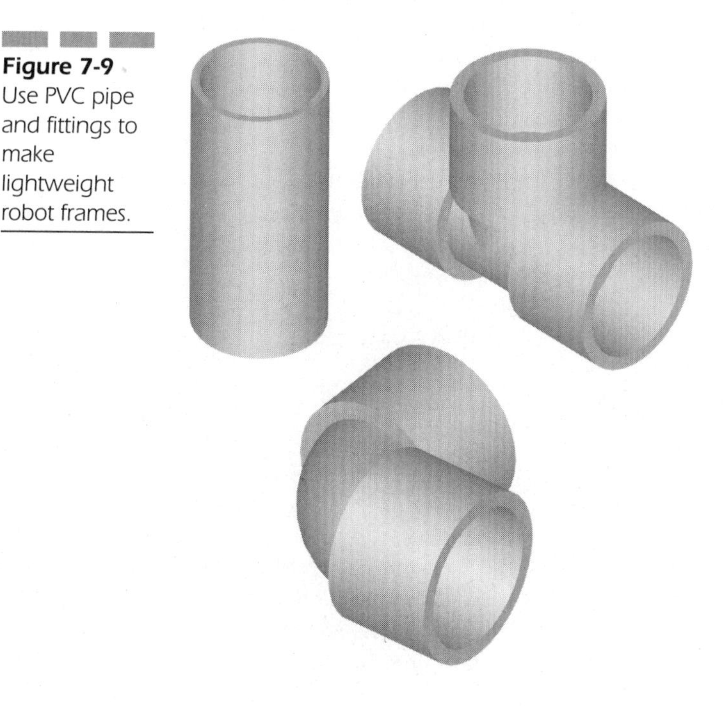

Figure 7-9 Use PVC pipe and fittings to make lightweight robot frames.

The pipe is made to fit snugly into the connectors. PVC is easy to cut and drill, so you can readily attach other components to the frame. However, do consider that the PVC plastic used in these fixtures is rather soft and flexible, so it's limited to lightweight robots.

Yet another idea is to use so-called furniture-grade PVC pipe for small robot frames. This pipe, available in sizes ranging from 3/4" to 2", is intended to make outdoor furniture using PVC pipe and fittings. It is furniture grade because the pipe has been treated with an ultraviolet-resistant chemical (ordinary PVC becomes brittle over time from exposure to ultraviolet light).

As with retail store fixture materials, you can cut the PVC pipe to length and assemble it using various connectors. Use PVC solvent cement to prevent the pipe from working loose (or you can use self-tapping screws if you want to be able to disassemble the frame in the future). Cut holes in the pipe to mount other components.

Plastic Extrusions

Although aluminum is a popular material for machine framing, it is not your only choice. On one extreme are steel frames and structural shapes (e.g., tubes, squares, rectangles). These are quite heavy, much harder to cut, and need painting or other coating to inhibit rust.

A better approach for the typical amateur robot is plastic extrusion (Figure 7-10), where the plastic is styrene, PVC (and PVC pipe, as mentioned previously), ABS, or a composite. Plastic extrusion is not as strong as aluminum, and supplementary hardware is not as extensive. However, it costs considerably less. Like aluminum extrusions, the common style in plastic is the T-slot, either with a square shape or an octagonal profile.

Frames are built with plastic extrusion using special slip-on connectors. The extrusion slides over the connector for a fairly snug fit. Additional structural strength can be achieved by drilling holes into the plastic framing and connector, and then fastening with self-tapping screws.

Building Frames with Structural Shapes

Though T-slot profiles (aluminum or plastic) are among the most common shapes for machine framing, you can also use a number of structural shapes for building sturdy, yet lightweight, robot bases. Round, square, and rectangular tubing are among the most common structural framing, and their cost is reasonable. Figures 7-10 through 7-16 are a partial sample of what's available.

Figure 7-10 Plastic extrusions are cheaper than aluminum but not quite as strong.

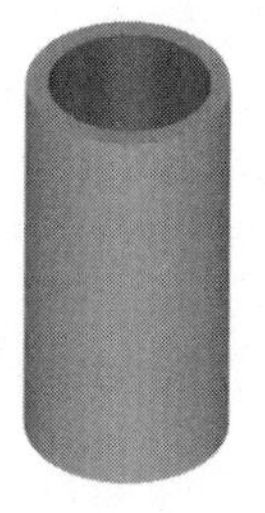

Figure 7-11

Round Tube (or Pipe)

Use a round tube with pipe or specialty connectors.

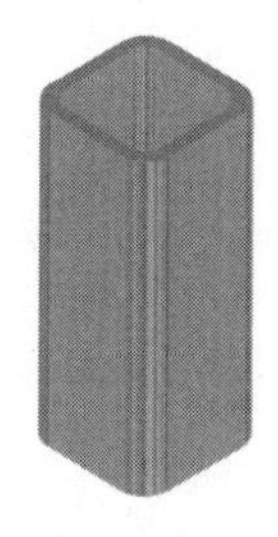

Figure 7-12

Square Tube

Use a square tube with specialty connectors or with standard bracket hardware to assemble frames. Available thicknesses are limited.

Figure 7-13

Rectangular Tube

Use a rectangular tube in the same manner as you would use a square tube, but it has a rectangular profile. Use with standard bracket hardware to assemble frames. Available thicknesses are limited.

Figure 7-14

Angle (Equal and Unequal)

Angle tubes are available with equal and unequal sides. Frames can be constructed using standard bracket hardware, pop rivets, and machine screw fasteners. Available in thicknesses ranging from 1/32" to over 1/4".

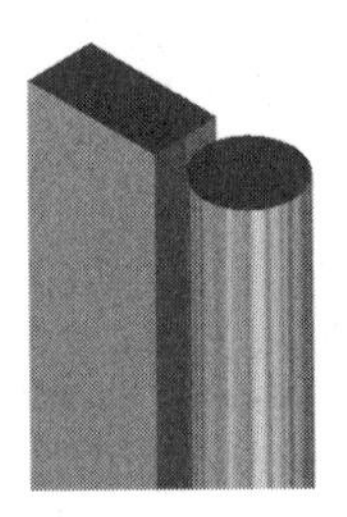

Figure 7-15

U-channel (Even and Uneven)

As with angle extrusions, U-channel tubes are available with equal and unequal dimensions, and use the same construction. Available in thicknesses ranging from 1/16" to over 1/4".

Figure 7-16

Rod and Bar

Best used for structural support, rod and bar are available in thicknesses ranging from 1/16" to over 1/4".

These structural shapes are available in aluminum, brass, and plastic (common plastics are ABS, expanded foam PVC, *high-density polyethylene* [HDPE], and styrene). For strength, opt for aluminum. The brass is handy when you want metal but want a more rigid material than aluminum.

The following are good sources for structural shapes:

- **Larger-dimension shapes** These can be found in the local hardware and home improvement store. Look in the hardware section.
- **Smaller-dimension shapes** These can be found in hobby and craft stores. Some hardware and home improvement stores also carry a limited assortment of the smaller pieces. A popular maker of metal structural components is K&S Engineering. Check their Web site at ***www.ksmetals.com*** for a description of their products.

Experimenting with *Cutless* Metal Platform Designs

A *cutlass* is a very sharp sword, principally used by pirates, to mince up people into tiny pieces. A *cutless* (my word) is a kind of robot platform built without any saw cuts. The idea of the cutless is not only that it's easier to build, but that you escape being minced up into tiny pieces by a wayward saw. The part about you being a pirate is optional.

The basic idea behind the cutless is to use base materials that are already the proper size and shape. The parts of the robot—the motors, sensors, batteries, and so on—can then be attached using fasteners, glue, hook and loop, double-sided foam tape, tie-wraps, and other techniques. The concept seems simple (and obvious) enough, but it's amazing how difficult it can be to find ready-made things that are exactly the right size, shape, and strength for your robot.

That is, until you learn how to spot the better candidates. The trick to the cutless method is to think outside the box and visualize everyday things used for robot bases. A prime source for materials for cutless bases is the hardware store, but other outlets shouldn't be ignored. Keep your eyes open and you'll note many ready-made components that can be used, without any additional sawing or sanding, for a robot base.

Following is an example of a cutless mobile robot design using commonly available (and inexpensively priced) metal pieces. Metal is a handy material because it's compact yet strong.

Mini T-bot

The Mini T-bot is made from a 6" strapping T (also spelled tee), commonly used in lashing together pieces of lumber in a home. Strapping Ts are available in numerous sizes; the 6" size is the smallest, but they are also available up to 16". The size measures the top of the T; the vertical portion of the T is in various lengths, depending on its design.

One popular strapping T is the Simpson Strong-Tie T Strap; I purchased mine at the local Home Depot, but they are sold at many local and online hardware stores. The brand doesn't matter; anything similar will do. The Mini T-bot uses the Simpson 66T, made of 14 gauge galvanized steel, and measuring 6" × 5", with a strap width of 1½". The 66T, like most strapping Ts, has holes in it for nailing. The holes are offset and most will not line up with the kind of hardware we want to hang on a robot, so we'll need to drill new holes. A power drill or better yet a drill press is recommended for drilling the holes.

Mini T-Bot Template The basic layout template for the Mini T-bot is shown in Figure 7-17. The robot uses the following parts, in addition to the strapping T and assorted fastening hardware:

- Tamiya worm gear motors, model 72004 (two)
- Tamiya narrow tires, 58 mm diameter, model 70145 (one set of two tires)
- Tamiya ball caster, model 70144 (comes in sets of two, only one used)

The Tamiya parts can be purchased from TowerHobbies.com, as well as from similar online retailers. The motors are mounted on the ends of the cross, and the

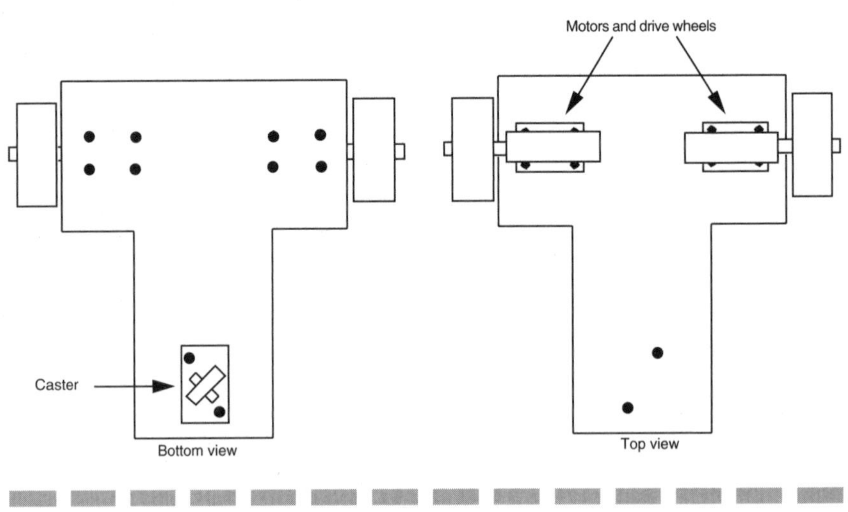

Figure 7-17 The Mini T-bot requires no cutting.

caster is mounted at the base. The Tamiya caster offers the option of two heights; use the longer height to better match the wheelbase afforded by the motors and tires.

Only a few holes need to be drilled in the strapping T. I used a 5/16" drill and 4-40 by 1/4" hardware; the small fasteners and the larger holes provide some slop in mounting. The slop allows for aligning the caster (which is not critical) and the two motors (which is critical). Total weight of the Mini T-bot prototype, with 66T strapping T, motors, wheels, caster, battery holder, battery, 25-column breadboard, and assorted small switches, is 17.5 ounces (496 grams). Note that the four AA alkaline batteries alone contribute 3.5 ounces (about 100 grams) to the weight of the robot.

For your reference, the specifications of the most commonly available sizes of Simpson Strong-Tie strapping Ts and their weight in ounces and grams are listed in Table 7-5. Larger robots can be built using a bigger strapping T. Do note the weight: The 1212T strap weighs almost a pound and therefore needs motors that can haul that weight.

Going Further

Of course the concept of the cutless extends beyond the Mini T-bot. You can use the same idea for other robot designs made out of different metal materials. The key points to keep in mind are as follows:

- The material should already be in the size you need, so no cutting is required.
- Drilling may be needed. Avoid materials that already have lots of holes. The holes may not line up with the motors and other components you wish to add, and the existing holes can cause trouble when drilling new ones so close by.
- Avoid very thick materials for small robots, as they add unnecessary weight.

Table 7-5 Selected Simpson Strong-Tie strapping T's

Model	Material	L	H	W*	Weight
66T	14-gauge galv.	6"	5"	1 1/2"	5 oz; 142 gr.
128T	14-gauge galv.	12"	8"	2"	11 oz.; 312 gr
1212T	14-gauge galv.	12"	12"	2"	14 oz.; 397 gr.

*Width is the width of the strapping metal.

Figure 7-18 The buggy 'bot uses a cutless design based on a small aluminum sheet.

- Consider sheet materials that can be bent to create unusual robot base shapes. An example is shown in Figure 7-18. This buggy 'bot is made from an uncut 6" × 12" aluminum sheet purchased at the local hobby shop.

Working Tips

Nothing beats experience. Mistakes are a common by-product of gaining experience. In books like this, it's the writer's job to make as many observations and mistakes as possible, so he (or she) can then tell you about them. What follows is a short and concise collection of tips on working with wood, several kinds of plastics, and metals. In some cases, the tips are repeats of recommendations made earlier in this chapter and other chapters, but they are replicated here for the sake of completeness.

Of course, this collection of tips won't prevent you from making your own mistakes. They're inevitable as you work on your robots. Don't look at them as mistakes but opportunities to increase your skill.

Tips for General Safety

Working with tools involves some risk. Knife blades can break off and fly through the air. Chips of wood, plastic, or metal can be propelled into the eyes of unsuspecting onlookers. Fingers can stray too close to spinning saw blades. Safety is often taken for granted in the shop, and that's bad. Always work with safety in the front of your mind, not the back of your mind. You'll enjoy robotics so much more with all your body parts intact and functional.

- *Safety glasses or goggles* should be worn at *all* times, even during simple hardware assembly. Get a pair that is comfortable and provides an unobscured view. Safety glasses should also be worn by any spectators in the shop.
- *Ear protection* is highly recommended when using power tools such as saws or high-speed drills, and especially when using pneumatic tools.
- *Periodically inspect your tools* to ensure they are in proper working order and that all safety devices are functional. This is particularly important for saws. The guards should open easily and not jam as the saw cuts into the wood.
- *Never defeat the safety device of a tool*. Eventually the tool will defeat you.
- *Keep your cutting and sawing tools sharp*. Resharpen or replace dull tools. Often, the most common and serious accidents occur because a saw or drill bit is dull, and the operator applies too much force on the tool.
- *Use clamps and vises*. When using power tools, don't hold the work in your hands.
- *Never work barefoot*. It poses an electrical shock hazard, and you may step on sharp shards and scraps of material.
- *Don't wear loose-fitting clothing or jewelry*. Roll up the sleeves of your shirt and remove your tie or tuck it into your shirt.
- *A shop apron* will keep your clothes clean. The apron should not tie in the front.
- *Work only in well-ventilated areas*, especially when applying paints or adhesives or when soldering, brazing, or welding.

Tips for Working with Wood

Of all the materials used in robot making, wood is the easiest to work with. Yet wood is not without its limitations and troubles. Keep in mind the following:

- *Wood can warp if it is exposed to moisture*. The warpage can occur when the wood swells with moisture or when it dries back out. Warpage is difficult to remove, so it's best to avoid it in the first place. Keep your wood stock in a cool, dry place, and once you build something with it, paint or seal the wood to prevent warpage later on.
- *Use only sharp saws and drill bits*. Dull tools make you work harder, and the extra friction can burn the wood.

- *Cutting with the grain will go faster* than cutting across the grain. Alter the feed rate accordingly. Slow down when cutting across the grain.
- *When drilling, back the wood with a piece of scrap*. This helps prevent splintering as the bit punches through. You may also place a piece of masking tape on the entry and exit points of the wood. This also helps reduce splintering.
- *Avoid bearing down hard when drilling*. This only serves to dull the tip of the bit. Work slowly. Let the tool do most of the work.
- *To drill large holes, start with a smaller hole*.
- *For deeper wood, periodically lift the bit out* of the hole to remove the built-up saw dust and wood chips.
- *Hardwoods are recommended* for robot bases and frames over softwoods, but stay away from the very heavy hardwoods such as oak. Though oak is plentiful and fairly inexpensive, it adds too much weight to your robot.

Tips for Working with Expanded PVC

Expanded (or foamed) PVC is one of the best materials for constructing strong robot bodies. It's relatively cheap, lightweight, and easy to work with. The material is available in sheet form, as well as in tubes, squares, rods, and other shapes. The following are some details about working with expanded PVC:

- *Cut and drill PVC slowly*. Although foamed PVC can be sawed and drilled like wood, the feed rate should be slower. This prevents gumming up the drill or saw.
- *During cutting or drilling, cool the cut with a blast of compressed air*. A flexible hose can be set up on your saw or drill press for this. The air also removes chips that can gum up the saw or drill.
- *You can use most any high-speed cutting tools with expanded PVC*. Band saws, circular saws, and panel saws are good choices. I use a radial arm saw, equipped with a carbide-tipped blade, to cut out pieces of PVC sheets.
- *For circular saws, blades with a minimum of 80 teeth are recommended*.
- *When using a reciprocating saw, use blades with a tooth pitch of 0.080" to 0.32"* to give best results, but remelting of the plastic into the cut (because of friction) is always a problem.
- *Wood, metal, or plastic drills can be used on expanded PVC*. If using milling tools, ground them the same as you would when using other plastics. Drills should have a 30-degree angle of twist in order to facilitate chip removal, or else the plastic may remelt into the cut.
- *For faster cutting, scoring, punching, or perforating, warm PVC to 85 to 105 degrees* (an incandescent or infrared heat lamp is a good choice).

- *Deep grooves or notches should be avoided*, as this reduces the strength of the material.
- *Solvent-based materials* containing *methyl ethyl ketone* (MEK) or *tetra hyrdo furan* (THF) are often used to join PVC pieces together. Both of these chemicals *are highly toxic*, however. Treat them with respect. These chemicals are the most common solvent ingredients found in PVC cements sold at hardware and home improvement stores.
- *Apply the solvent with a narrow artist's brush*, or better yet, use a needle-point applicator. For the latter, you need low-viscosity solvent, such as WELD-ON 2007.
- *Clean all surfaces to be joined with isopropyl alcohol.*
- *Never use an open flame.* Expanded PVC is thermoformable, so it can be heated and bent into shapes. Use a heat gun, heat strip, or other apparatus approved for use with bending plastics. Never, ever use an open flame to heat PVC, or any other plastic. Though PVC is flame retardant, it can catch on fire; if it does, it releases toxic gas, and the molten plastic can drip and cause very serious burns.

Tips for Drilling and Cutting Acrylic

All things considered, acrylic plastic isn't the best material for building robots, but it's relatively cheap and can be found everywhere. You're better off using expanded PVC plastic, or even ABS.

But ABS and PVC plastic sheets can be hard to find. Sometimes, we have to use what we can get, so if you're stuck building a robot out of acrylic plastic, here are 10 tips for drilling and cutting it into the shape you need.

- *First and foremost, always wear protective goggles or safety glasses*. Acrylic can shatter when cutting or drilling it, and pieces may fly off and lodge in your eye. Use caution—plastic can be nasty!
- *Avoid using wood drill bits*. Drill bits designed for plastic or glass drilling yields better, safer results. You probably need only one or two plastic bits; 1/8" and 3/16" sizes should suffice for most work.
- *Use a power drill, not a hand drill*. You get best results by using a medium drill speed: about 750 to 1,000 rpm. (For wood you want a fast speed, for plastic a medium speed, and for metal a slow speed.)
- *When drilling, always back the plastic with a wooden block or a piece of masking tape*. Without the block or tape, the plastic is almost guaranteed to crack.
- *Holes larger than 1/4" should first be made by drilling a smaller pilot hole*. Start with a small drill and work your way up several steps.

- *Use a radial arm or table saw for straight cuts*. Though there are specialty blades for cutting plastics, they are expensive. A regular combination blade is acceptable for all but the thinnest of acrylic sheets.
- *Use only a sharp blade*.
- *Avoid fast feed rates* and don't force the material into the blade. When cutting (and drilling for that matter), the plastic tends to remelt at the cut, and pushing the work too fast can generate lots of friction.
- *For rounded cuts, use a saber saw or scroll saw* (the latter is my preference). Or, if you don't have to cut much, use a manual coping saw. Use a slow speed to avoid remelting the cut plastic due to friction.
- *Finish the cut ends and sharp corners of the plastic with a metal file*. If you have the equipment, you can also burnish the edges with plastic rouge and a buffing wheel. The experts use flame burnishing to obtain smooth edges, but I don't recommend it if you're not skilled in the art.

Tips for Working with Metal

The following are some tips to streamline your metalworking:

- *Always use sharp, well-made drill bits, saws, and files*. Dull, bargain-basement tools aren't worth the trouble.
- *If you're using a power drill or power saw, use the slower speed settings*. High-speed settings are for wood; slow are for metal.
- For sawing metal, select a fine-tooth blade, on the order of 24 or 32 tpi. Coping saws, keyhole saws, and other hand saws are generally engineered for woodcutting, and their blades aren't fine enough for metal work.
- *When cutting rod, bar, and channel stock by hand, use a miter box*. The hardened plastic and metal boxes are the best buys. Be sure to get a miter box that lets you cut at 45 degrees both vertically and horizontally.
- *Use a punch to ensure accurate drilling*. When cutting metal, the bit will skate over the surface until the hole is started. The punch creates a small dimple in the metal that reduces skating.
- *Whenever possible, use a drill press to cut metal pieces*.
- *Always use a proper vice when working with a drill press*. *Never* hold the work with your hands, or serious injury could result. If you can't place the work in the vise, use a pair of vice grips or other suitable locking pliers.
- *Cutting and drilling leaves rough edges* (called burrs and flashing) in the metal. These must be filed down using a medium- or fine-pitch metal file.

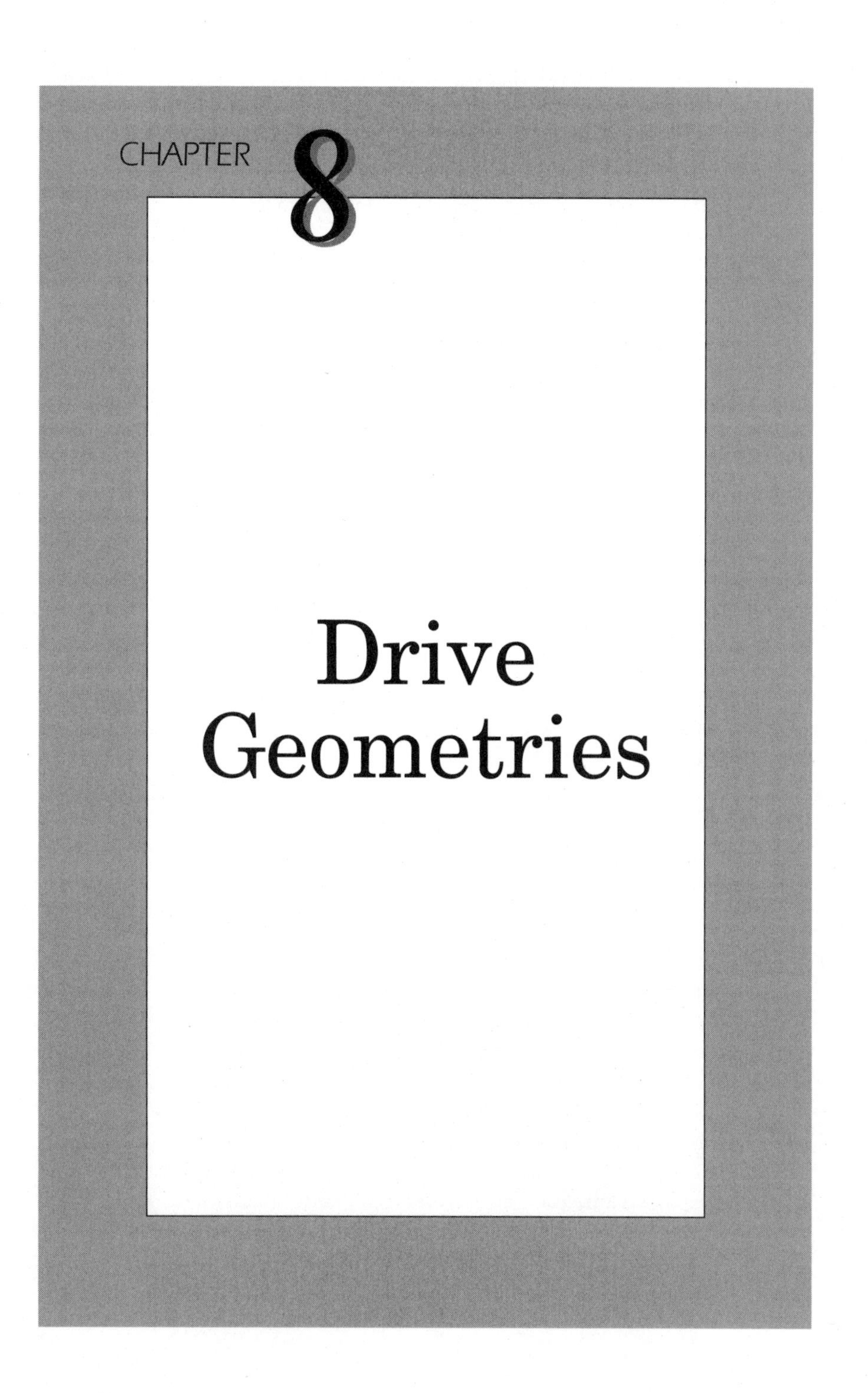

CHAPTER 8

Drive Geometries

All mobile robots have one thing in common: wheels, tracks, legs, or other means to propel them across the floor. How these wheels, tracks, and legs are arranged is called the *drive geometry*. There are many variations of drive geometries, each with a set of pros and cons.

Selecting a drive geometry involves not only analyzing the desired functionality of your robot, but assessing the mechanical requirements of constructing the drive mechanism. Such assessments would include understanding that tracked and legged bases are harder to build than wheeled robots, and that legged mechanisms made with certain plastics (primarily acrylic) can break over time because of stress.

In this chapter, you'll learn about drive geometries and how to select the best one for your application. Because wheeled and tracked robots are by far the most common variety and represent the biggest choices, we'll spend the majority of our time with them.

Choosing a Locomotion System

In previous chapters, we discussed the major variations of robot locomotion: wheels, tracks, and legs. The cogent points are worth repeating in Table 8-1, and we'll raise a few new important issues as well.

Drive Arrangements of Wheeled Robots

With few exceptions, cars have four wheels. They're predictable and everyone knows what to expect. Not so with mobile robots. Wheel arrangements can vary, with some robots having only two wheels (okay, a few even have only one wheel!), and others having six and even eight. The wheels may pivot to make a turn, or they may remain fixed to the frame; the robot negotiates a course by selectively powering one wheel or another.

Drive arrangements can be clearly defined, and the different arrangements provide specific advantages. The number and arrangement of the wheels also determine how the base is steered. On some robots, steering is accomplished by pivoting one or more wheels; on others, the vehicle is steered by altering the speed and direction of wheels.

Table 8-1 Considerations of robot locomotions

Locomotion	Drive Considerations	Mechanical Considerations
Wheels	The most common arrangement is two wheels on opposite sides of the base, with one or two skids or casters for balance. Common variations include four- and six-wheeled bases, which do not require balancing casters or skids. The size of wheel greatly influences the traveling speed of the robot. Larger wheels (for a given motor's rpm) make the robot go faster. On two-wheeled robots with a support skid or caster, the wheels can be mounted at the center line in the base, or they can be offset to the front or back. Distance measuring (*odometry*) is more reliable with wheeled bases. Accurate calculations are more difficult with tracked and legged robots.	Mounting wheels to motors or wheels to a shaft is the hardest part of building a wheeled base. *Radio control* (RC) servo motors provide a consistent means for mounting small wheels to them, so these types of motors are quite common in small mobile robots. A modest degree of accuracy is needed in mounting the wheels to avoid *run-out*, side-to-side wobble as the wheel rotates.
Tracks	The treads form a wide base that enhances the stability of the vehicle. The mechanics of the treads creates a virtual wheel with a very large surface area that contacts the ground. There's no need for a support caster or skid. Though not as common, the treads may be augmented by wheels—similar to the half-track military vehicle. The treads are shorter and support only one end of the base.	Suitable tread material can be hard to find; the most common approach is to hack a toy tank. The large surface area of treads greatly increases friction; tracked vehicles can have trouble making turns, and the treads can pop off. Rubber treads (the most common on hacked toys) can stretch over time. A track tensioner mechanism is recommended.
Legs	Variations include two, four, six, and even eight legs; six legs (*hexapod*) are the most common.	Of all locomotion types, this one requires the greatest degree of machining and assembly.

Table 8-1 Considerations of robot locomotions (Continued)

Locomotion	Drive Considerations	Mechanical Considerations
	Most legged robots use *static* balance, meaning the arrangement of the legs on either side of the base prevents it from toppling over. Much rarer is *dynamic* balance, where weight on the base is shifted to compensate for stepping.	The flexing of legs can cause stress in material; acrylic plastics can break over time.
	The joints of each leg are defined as *degrees of freedom* (DOF); the more DOF, the more agile the platform, but the more difficult to build.	Legs with independent articulation (each leg can move separately and independently) is the most difficult to construct. An easier alternative is the linked-gait articulation, where the movement of legs are linked together. Fewer moving parts and motors are required.

Despite the variety of geometries, robot drives can be placed into two distinct categories: holonomic and nonholonomic. These are described in the following section.

Holonomic Versus Nonholonomic Drive Systems

Technically speaking, *holonomic constraints* refer to the relationship between coordinates describing both position and orientation. This is better described by example. A car is a nonholonomic vehicle. It cannot move in any direction instantaneously. It can travel forward or backward, and it can veer to the left or right. The orientation of the wheels of the car prohibits it from instantaneously moving in any direction of the compass.

Holonomic drives are distinctive in that they have the same number of controllable *degrees of freedom* (DOF) as their actual number of DOF. Again, this is better described by example: A ball is a simple holonomic entity. It can roll in any direction, from any direction, at any time. It can instantaneously travel straight, and then move in any direction of the compass.

Coaxial Drive The most common nonholonomic system in robotics is the *coaxial drive*, which in its most basic form consists of two wheels mounted on either side of the robot, as shown in Figure 8-1. It's called coaxial because the two wheels share a common axis, though in practical use, the wheels don't have to

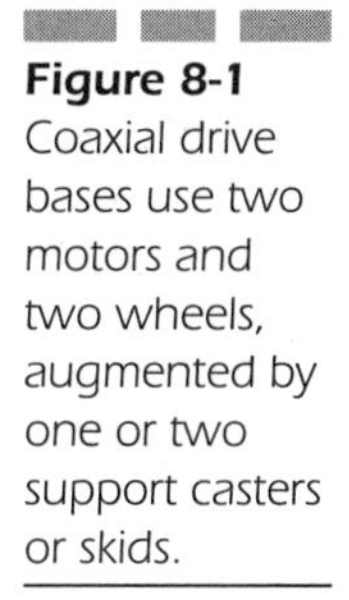

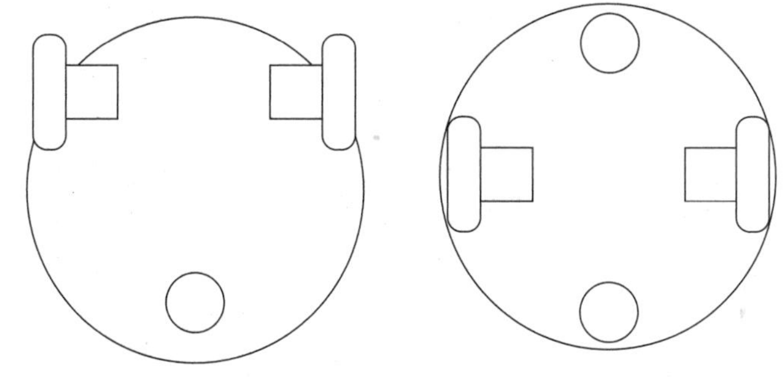

Figure 8-1 Coaxial drive bases use two motors and two wheels, augmented by one or two support casters or skids.

be exactly opposite one another. One or two casters or skids, placed at the center line over the robot in the front or back, provide support for the base.

Variations of the two-wheeled base include four, six, and even eight wheels. On bases that use more than two wheels, support casters or skids are not needed. One technique is the dually drive, where a single motor drives two wheels that are placed close to one another (see Figure 8-2). Each of the four wheels can be driven independently by separate motors, or the wheels on each side can be driven by a single motor, linked via a belt, chain, or gear system.

One of the key benefits of coaxial drive is that the robot can spin in place by reversing one wheel relative to the other, as shown in Figure 8-3.

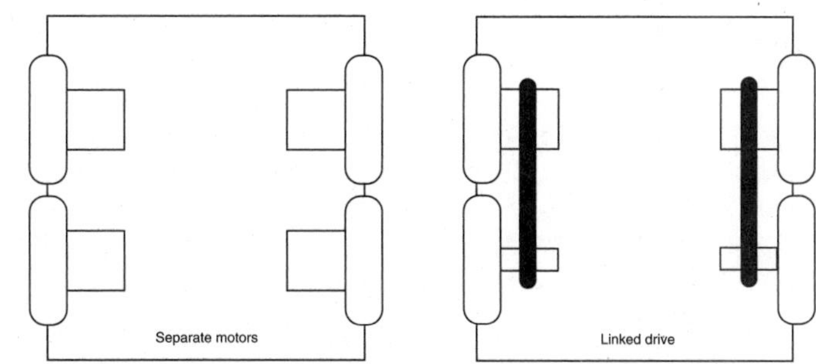

Figure 8-2 4WD robots use four wheels. In one approach, each wheel is driven independently; in another, the wheels on each side are linked via a chain, belt, or other form of power transmission.

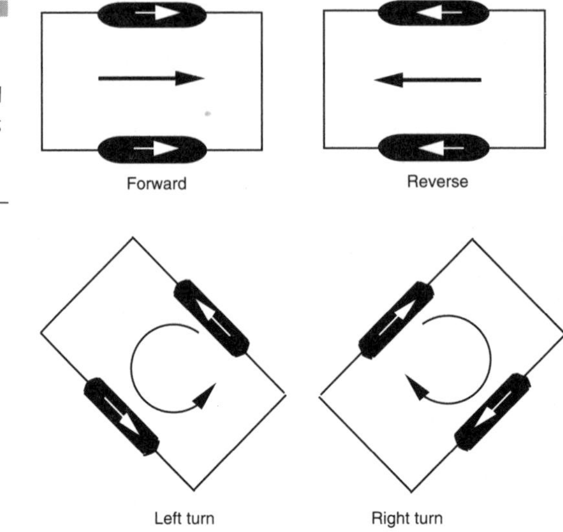

Figure 8-3 Coaxial drive and differential steering allows a robot to spin in place.

NOTE: *I had to make up the term "coaxial drive" because there isn't a good descriptor for this type of wheel arrangement. It's also referred to as differential steering—and inaccurately as differential drive—but I wanted a term that defined the relationship of the drive mechanisms, not just the steering. The two are actually distinct.*

Car-Type Steering Pivoting the wheels in the front is yet another method of steering a robot. Robots with *car-type* steering (see Figure 8-4) are not as maneuverable as differentially steered robots, but they are better suited for outdoor use, especially over rough terrain.

Somewhat better traction and steering accuracy are obtained if the wheel on the inside of the turn pivots to a greater extent than the wheel on the outside. This technique is called *Ackerman steering* and is found on most cars but not on many robots. Ackerman steering is preferred for larger vehicles because the turn produces more traction and less wear on the tires. As shown in Figure 8-5, the wheel on the inside of the turn describes a smaller circle than the wheel on the outside.

Three-Wheel Tricycle Car-type steering, described previously, is one method that avoids the problem of crabbing due to differences in motor speed (simply because the robot is driven by only one motor). But car-type steering creates fairly cumbersome indoor mobile robots. A better approach is to use a single drive motor powering two rear wheels and a single wheel for steering in the front; the arrangement is like a child's tricycle (Figure 8-6). The robot can be steered in a circle slightly larger than the width of the machine. Be careful of the wheel base

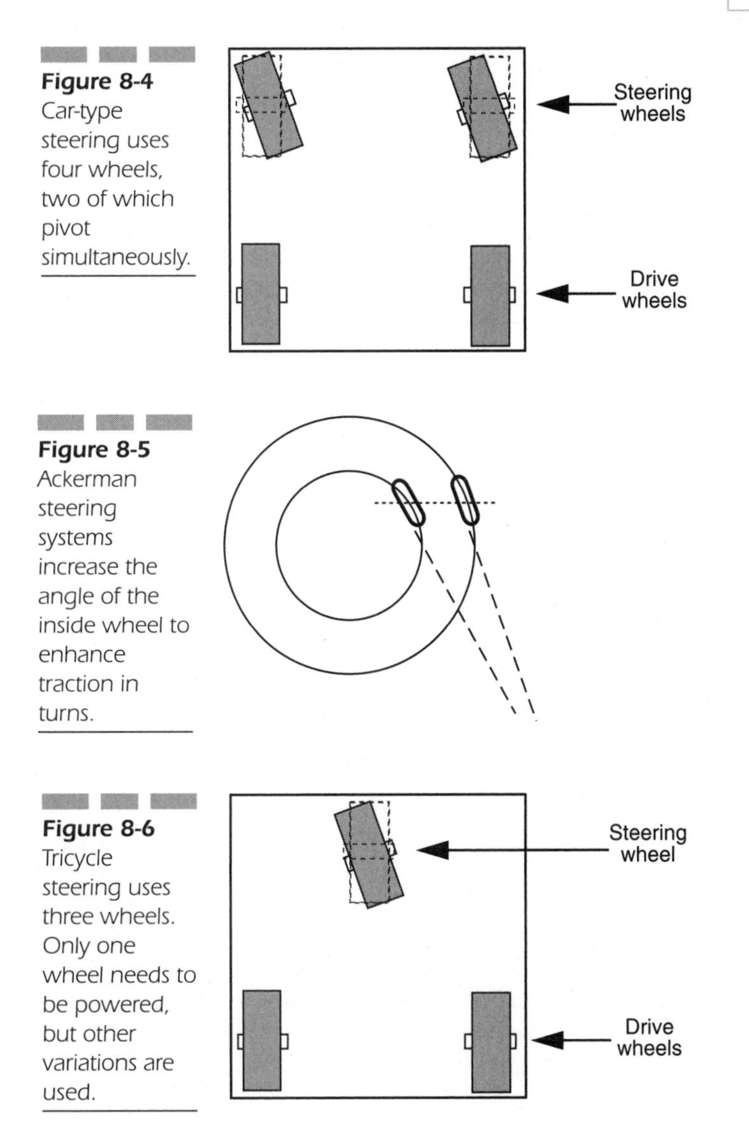

Figure 8-4 Car-type steering uses four wheels, two of which pivot simultaneously.

Figure 8-5 Ackerman steering systems increase the angle of the inside wheel to enhance traction in turns.

Figure 8-6 Tricycle steering uses three wheels. Only one wheel needs to be powered, but other variations are used.

of the robot (i.e., the distance from the back wheels to the front steering wheel). A short base will cause instability in turns, allowing the robot to tip over opposite the direction of the turn.

Tricycle-steered robots require a very accurate steering motor in the front. The motor must be able to position the front wheel with subdegree accuracy. Otherwise, there is no guarantee the robot will be able to travel a straight line. Most often, the steering wheel is controlled by a servo motor; servo motors use a closed-loop feedback system that provides a high degree of positional accuracy.

There are two basic variations of tricycle drives:

- **Unpowered steered wheel** The steering wheel pivots but is not powered. Drive for the robot is provided by one or two other wheels.
- **Powered steered wheel** The steering wheel is also powered. The two other wheels freely rotate only.

A subvariant of the tricycle-based design reverses the functionality of the wheels: Two wheels in the front of the robot steer, and a third wheel in the back provides support. The third wheel can even be a simple caster or omnidirectional ball (see the section on caster types later in the chapter).

Holonomic Bases

On the holonomic end of the spectrum, the common trait of all the drive systems is that the robot is omnidirectional, able to move in both the X- and Y-planes with complete freedom. A common form of holonomic base uses three motors and wheels, arranged in a triangle configuration (some bases use four motors and four wheels). Figure 8-7 shows two possible base styles, which can be round, triangular, or any shape.

In this type of holonomic base, the wheels have rollers around their circumference; they provide traction at angles other than perpendicular to the hub of the wheel. The robot moves forward by activating any two motors, and it turns by adjusting the speed or direction of any one or all three of the motors. These types of wheels were originally designed for the handling of materials as a substitute for conveyor belts.

Perhaps the most famous robot that uses this design approach is the *Palm Pilot Robot Kit* (PPRK). It uses a Palm Pilot for its controller, hence its name. The robot was designed at the Robotics Institute at Carnegie Mellon University and is available in kit form from several vendors. More information on this robot can be found at ***www-2.cs.cmu.edu/~pprk/***.

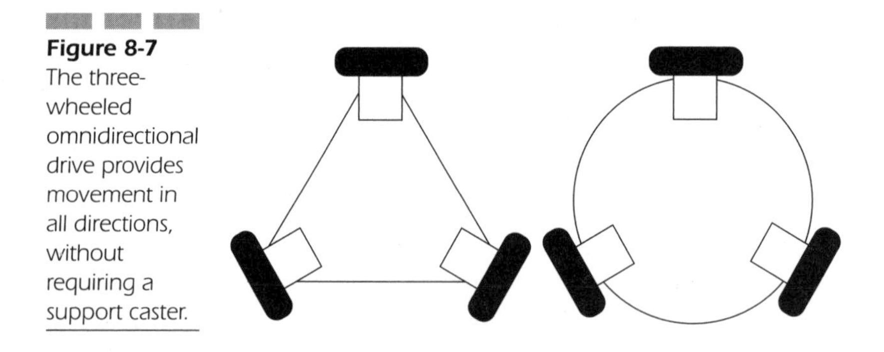

Figure 8-7 The three-wheeled omnidirectional drive provides movement in all directions, without requiring a support caster.

As you can imagine, this system requires three drive motors instead of merely two. An alternative design uses four wheels, with either two motors (two wheels per motor) for four motors. The wheels are mounted in traditional car fashion. The robot is differentially steered, as explained previously.

Other forms of holonomic bases involve drive wheels that can be rotated independently. These go by various names, such as synchronized omnidirectional. The rotation of each drive wheel can be accomplished by separate steering motors or by one steering motor that is linked—via a chain or cogged belt—to all of the drive wheels. An example is shown in Figure 8-8.

Tracked

Tracked robots are basically robots with coaxial drives and differential steering but without a support caster or skid. The track acts as a kind of giant wheel set, with one wheel on each side. If both tracks move in the same direction, the robot is propelled in a straight line forward or backward. If one track is reversed, the robot turns. The main benefit of a tracked vehicle is its ability to navigate over rough terrain. The tracks enhance the grip of the road, allowing the robot to travel over loose dirt, sand, grass, and other surfaces that a wheeled robot can only dream about.

***NOTE:** In a coaxial drive base, it is possible to turn by simply stopping one wheel; however, this is not advisable with a track drive. The reason: The track exposes a very large surface area to the ground. This enhances traction but also increases friction. The stopped track will skitter over the ground, and the amount of turning is harder to control.*

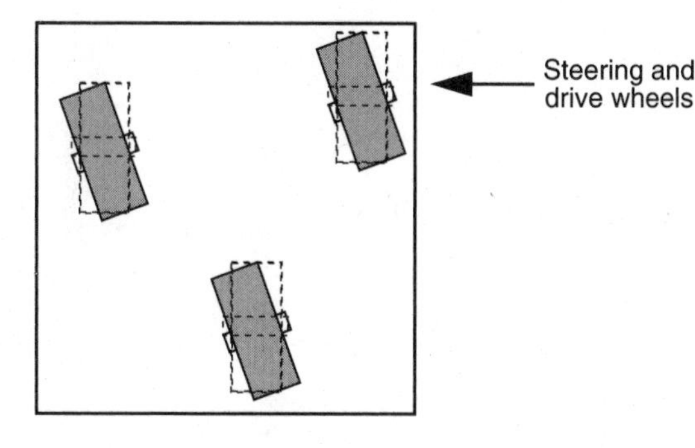

Figure 8-8 Synchronized omnidirectional bases are among the most difficult to build, but they deliver a robot that can move in any direction without changing its heading.

A variation of the tracked vehicle is the 6WD, or *six-wheel drive*. Both the traditional track and 6WD arrangements are shown in Figure 8-9. In both, a single motor on each side of the base propels a driver. For a tracked base, the track itself acts as a belt to drive the other support wheels. For a 6WD base, a belt, gear system, or chain drives all the wheels simultaneously. Movement and steering with the 6WD base are the same as with a tracked vehicle, except that traction and friction are somewhat less. One advantage to 6WD bases is that their drive train is not as exposed and, therefore, not as likely to become gunked up with dirt, grass, and other impediments.

Selecting a Drive Type

To recap:

Nonholonomic

- Coaxial, with one or two casters or skids
- Coaxial, four or more wheeled
- Tricycle
- Car-type or Ackerman
- Tracked or 6WD

Holonomic

- Three- and four-wheel omnidirectional
- Synchronized omnidirectional

Which one is better? Neither and both. The decision about which drive type to use relies on your plans for the robot, your budget, and your construction skills. The simpler two-wheeled bases are the easiest and least expensive to construct. If you're relatively new to robots, this type should be your first attempt.

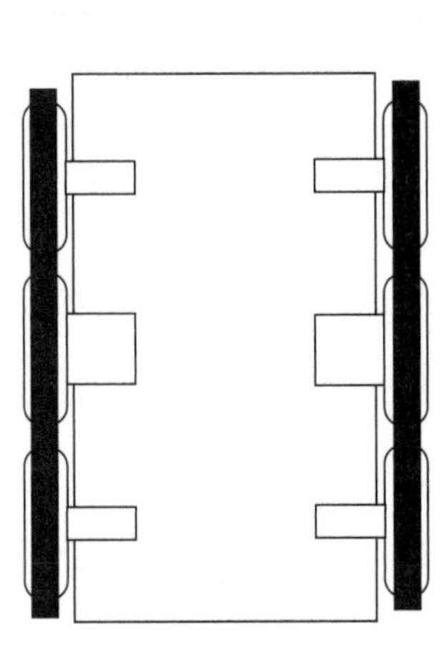

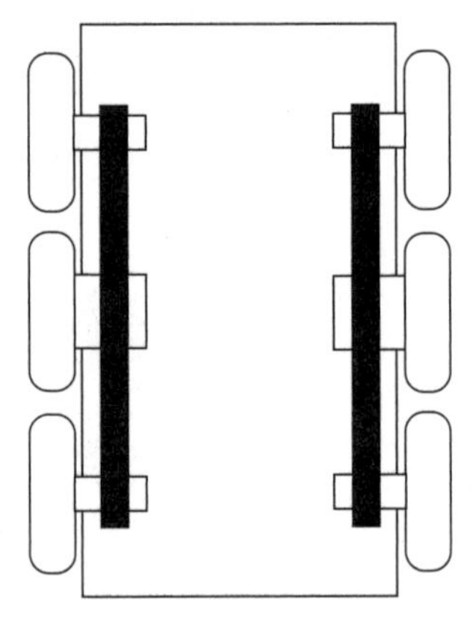

Figure 8-9 Tracked and 6WD bases are remarkable in their ability to travel over rough terrain.

Tracked and 6WD robots are more difficult to construct unless you use a commercially made toy or vehicle. For tracked vehicles, the choice of tread is fairly limited—most toys use flexible rubber treads. If your designs are for a larger and heavier-duty robot, you'll need to either make your own treads or purchase them ready made. Snowmobile treads are an option for very large robots. For smaller robots, 1/8- and 1/12-scale metal treads built for high-end tank models are a viable choice. These can be frightfully expensive, however; plan on spending $300 or more for a good set.

Holonomic robots that use roller-wheels (e.g., the PPRK) are only modestly more difficult to construct than standard two-wheel robots, but their control programming is significantly more complex. It's easy to run a two-motor robot: Simply turn the motors on to go forward, and stop or reverse one to make a turn. . Operating and turning a holonomic robot requires some higher-level math functions that not all builders are comfortable with. Additionally, the math functions may be too complex for the simple microcontrollers often found on amateur mobile robots, such as the Basic Stamp. Attempt this style of base only if you're also comfortable with the electronics and programming that go with it.

Choosing Wheels

There is no single type of wheel that works for every robot. Beyond obvious differences such as diameter and width, wheels differ in many ways, such as how they are mounted to a motor or axle, their hardness, and many other factors.

Wheels are made up of tires (or tyres, in the U.K.), mounted on hubs (Figure 8-10). A tire is rubber, plastic, metal, or some other material, and the hub is the portion that attaches to the shaft of the axle or motor. Similarly, the wheels on a car are made of the tire on the outside, and the hub, or rim, on the inside.

Some wheels for robots are molded into one piece. Others, such as the Dave Brown Lite-Flight wheels, are composed of two separate pieces assembled at the plant. The Lite-Flight wheels use a plastic hub that attaches to the motor shaft or axle, and a foam tire is mounted onto the hub.

Figure 8-10 Wheels are tires mounted on hubs.

Here are some basic factoids regarding wheels. If you want to learn more, be sure to read *Building Drive Trains*, one of the companion books in the Robot DNA series. It fully describes selecting and using wheels in mobile robots, including various ways of mounting them to motors.

Wheel Materials

The first order of consideration is what materials to use for the wheel. The least expensive wheels, such as those used on low-cost toys, are one molded piece, usually a hard plastic. These wheels don't have a separate tire and hub. Although these wheels are acceptable for some robots, you probably want a more tractable tire surface. This requires a softer rubber or foam over a rigid hub. Hubs are commonly made of metal, styrene, or other plastic.

Figure 8-11

Rubber over Plastic

The hardness of the rubber greatly influences traction. One common measure of hardness is the Durometer, tested by a device of the same name. A Durometer of 60 is relatively soft and pliable, 75 is medium, and 90 to 95 is quite hard.

Figure 8-12

Rubber over Metal

Typical of wheels made for RC racing, these are heavier and more sturdy, and well suited to bigger robots. You can also get small rubber tires mounted on aluminum hubs. These are typically sold at hobby stores as tail wheels for model airplanes.

Figure 8-13

Foam over Plastic

Foam wheels are also a mainstay in the RC racing field. Like their rubber counterpart, hardness varies, from a fairly soft 45 Durometer, to over 65 Durometer.

Figure 8-14

Rubber or Foam over Spoked Wheels

As the size of the wheel increases, so does its weight. Spokes are used to reduce the weight of very large wheels. Smaller bicycle or wheelchair wheels are suitable for larger robots.

Figure 8-15

Pneumatic Wheels

Traditional foam and rubber tires are merely fitted over their hubs. In a pneumatic wheel, the tire is filled with air, which gives the wheel more bounce and even rigidity. Wheels for wheelbarrows and some wheelchairs are pneumatic. You will also find some pneumatic wheels for high-end RC race cars and buggies.

Figure 8-16

Airless Tires

Similar in concept to the pneumatic wheel, airless tires are hollow and filled not with air, but with a rubber or foam compound. They are common in wheelchairs and heavy-duty material-handling carts.

Wheel Diameter and Width

There are no standards among wheel sizes. They vary by their diameter, as well as their tread width (the tread is the plastic, foam, or rubber material that contacts the ground).

- The larger the diameter of the wheel, the faster the robot will travel for each revolution of the motor shaft. You can quickly calculate linear speed if you know the speed, in revolutions per minute or second, of the motor. Simply multiply the diameter of the wheel by pi, or 3.14, and then multiply that result by the speed of the motor. For example, if the motor turns at

1 revolution per second, and the wheel is 1" diameter, then the robot will travel at about 3.14" per second.

- The larger the diameter of the wheel, the lower the torque required of the motor. Wheels follow the laws of levers, fulcrums, and gears. As the diameter of the wheel increases, the amount of torque delivered by the wheel decreases.
- Wider wheels provide a greater contact area for the wheel, and therefore traction (from friction) is increased.
- Wider wheels tend to influence the *tracking* of the robot—tracking is the ability of the robot to travel a straight line when its drive is propelling it forward. Additionally, the wider tread area of the wheel increases friction, which acts to resist turns.

When selecting the wheel diameter and width, match the wheel to the job. A robot with modest-size wheels of fairly narrow proportions (say, 1/4" wide for a wheel of 2.5" to 3" in diameter) will be more agile than a robot equipped with much wider wheels.

Be mindful of the torque developed by the motors. For every doubling of the wheel diameter, torque decreases by approximately 50 percent. For example, if a motor that develops 50 ounce-inches (oz-in) torque, the torque will be half of that (or 25 oz-in) for a 2" diameter wheel, a third of that (about 16 oz-in) for a 3" diameter wheel, and so on.

Wheel Placement and Turning Circle

The position where the wheels are located on the robot base affects the turning circle, or radius, of the robot. Whenever possible, locate the wheels within the body of the base, rather than outside it. This decreases the effective size of the robot and allows the robot to turn in a tighter circle. Figure 8-17 shows wheels mounted both within the area of the base and outside it.

Note that when the wheels are outside, the effective diameter of the robot is increased. Assuming the wheels are mounted on the center line in the middle of the base, the turning circle of the robot is defined as the distance between the two wheels. For example, if the wheels are placed 8" apart, even if the robot base is smaller, the turning circle is 8".

Omnidirectional Wheels Go Your Way

Imagine a wheel that spins like any other wheel, yet also allows for sideways motion. That's an *omnidirectional wheel*, *multidirectional wheel*, or *omniwheel*—an idea that goes back to about 1910. The wheel is a series of small wheels or

Figure 8-17 Wheels mounted outside the area of the base increase the turning circle.

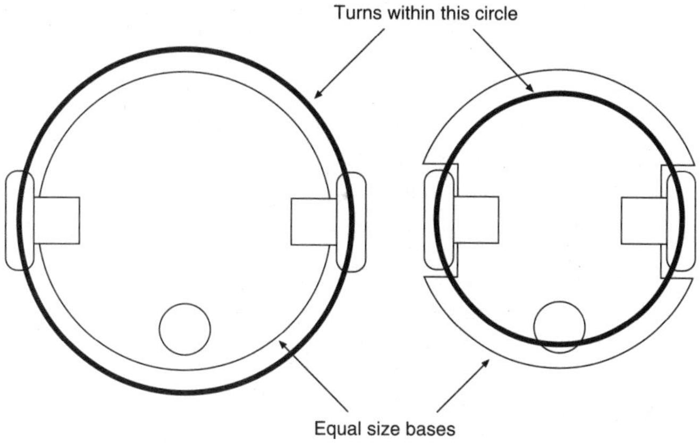

rollers, mounted around the circumference of a larger main wheel. They're popular in material-handling applications; the wheels are mounted in rows on top of tables or conveyors. Boxes or other goods glide effortlessly along the wheels and are allowed movement in any direction.

For robotics, omniwheels have two principle applications:

- Drive wheels are used in three-wheeled holonomic robots (see the earlier discussion on holonomic bases). An example of a robot that uses this design is the PPRK, sold by Acroname and several others. Rather than use two drive wheels positioned opposite one another, the PPRK uses three wheels in a triangular configuration. Only two motors propel the robot at a time, but the machine is able to move in any direction by applying power to specific motors.
- Freewheeling casters allow low-friction turning. A problem with swivel casters is that the swivel may not spin freely in turns. This causes the robot to lose accurate tracking and steering (if the robot is light enough, a caster that isn't pointed in the right direction will cause the little critter to veer off to one side!).

Omnidirectional wheels (see Figures 8-18a and 18-8b) are available in sizes ranging from about 40 mm (about 1.5") to over 150 mm (about 6"). The wheel material is acetal (hard) polyurethane or polyurethane. Alas, omnidirectional wheels are rather expensive, but for what they do, they do it well.

Here are some key sources for omniwheels, should you want to try some:

- Acroname, Inc. ***www.acroname.com***
- AIRTRAX, Inc. ***www.airtrax.com***
- Atlantic Conveying Equipment, Ltd. ***www.atlanticgb.co.uk***
- Budget Robotics ***www.budgetrobotics.com***

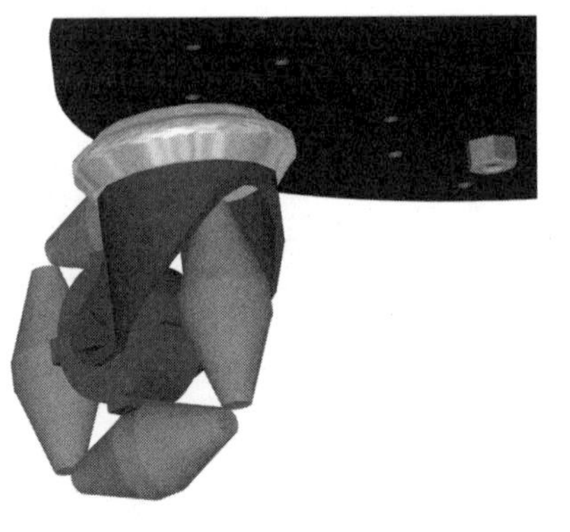

Figure 8-18a Omnidirectional wheels can be used as casters or as drive wheels.

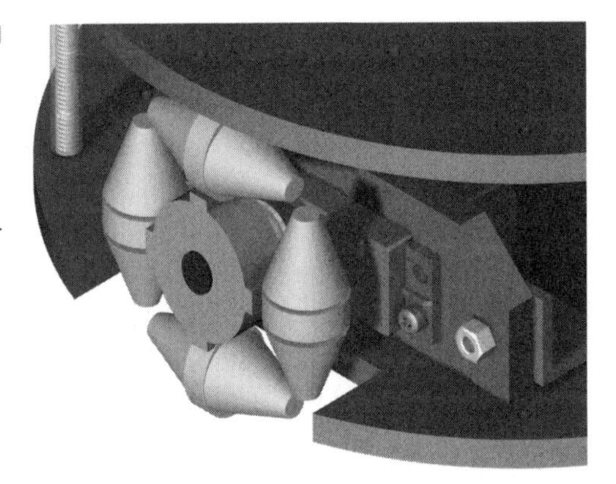

Figure 8-18b Another example of omnidirectional wheels.

- Kornylak Wheel Division ***www.omniwheel.com***
- Mr. Robot ***www.mrrobot.com***
- North American Roller Products ***www.narp-trapo.com***

Support Casters and Skids

Robots with coaxial drives—two drive wheels on either side—need something on the front or the back to prevent them from tipping over. Several common approaches are listed in this section.

If they were good enough for World War I airplanes, they're good enough for robots! Seriously, the purpose of skids, as shown in Figure 8-19, is to glide over the ground without using any moving parts. The skid is rounded to facilitate a

Figure 8-19 Nonrotating skid

smooth ride. Polished metal, hard plastic, and Teflon are common choices. Teflon skids are particularly low friction but are harder to find and more expensive.

For small lightweight robots rolling over low-nap carpet, wood, and tile, metal or plastic is sufficient. A cap nut secured to the end of a short machine screw is a perfect choice.

Skids are not suitable for robots that may travel over very uneven surfaces or may encounter many obstructions, such as cables and old socks. The skid may snag, and damage to the robot or its surroundings could result.

For robots measuring 8" to 10" or larger, or heavier than about 12 oz., swivel casters (see Figure 8-20) are recommended. Casters are available with wheel diameters from 1" to over 4"; match the size of the caster with the size of the robot. You'll find the common 1 5/8" to 2" diameter caster wheel is suitable for most medium robots (up to approximately 14"); use 3" and 4" casters for the larger and heavier brutes.

Swivel casters are commonly available with plate or stem mounting and in the following wheel styles:

- Single wheel
- Dual wheel (twin wheel)
- Ball

The ball style is commonly used with furniture and tends to be heavy for its size. Use it for larger and heavier robots. Single-wheel casters are probably the most common and easiest to find. Look for a caster that swivels easily. Twin wheel casters are ideally suited for lighter robots. The two wheels are made of

Figure 8-20 Swivel caster

Figure 8-21 Ball caster and ball transfer

lightweight plastic and are often a third or even half of the weight of similarly sized single-wheel casters.

Ball casters and ball transfers (see Figure 8-21) act as omnidirectional rollers. There's a difference between the two:

- Ball transfers are primarily designed to be used in processing materials. They're used with the ball pointing up in conveyor chutes and the like. Ball transfers are made of a single ball, either metal, rubber, or rubber held captive in a housing.
- Ball casters are similar, except they are designed to be used with the ball pointing down.

Some ball transfers can be used ball down. Those that cannot will exhibit impaired rolling and, eventually, permanent damage. When in doubt, ask the manufacturer or seller.

Unlike swivel casters, which must rotate to point in the direction of travel, ball transfers and ball casters are ready to move in any direction. This makes them ideal for use as support casters in robots.

The size of the balls varies from about 11/16" to over 3" in diameter, and they are available in steel, stainless steel, and plastic (primarily nylon). Look for ball casters at mechanical surplus stores, and also at industrial supply outlets, such as Grainger, McMaster-Carr, Outwater Plastics, MSC Industrial Direct, and others.

As noted above, omnidirectional wheels, shown in Figure 8-22, are basically rollers mounted on the tread of a tire. The tire turns on an axis like any other, but the rollers allow for movement in any direction.

Spherical Wheels, shown in Figure 8-23, are a unique, trademarked product. They're made as caster replacements for tables and chairs, but unlike regular casters, they are omnidirectional, and they do swivel. Functionally, they are quite similar to ball casters, but internally they use a different arrangement of rollers. They're also larger—sizes of the ball in Spherical Wheels range from 1" to over 4" in diameter. The smaller size is ideal for robots from 8" to about 16" in size; the 4" model can support over 100 lbs. and is suitable for larger 'bots.

Despite their obvious benefit to robots, there are few resources for Spherical Wheels. These include McMaster-Carr and Budget Robotics.

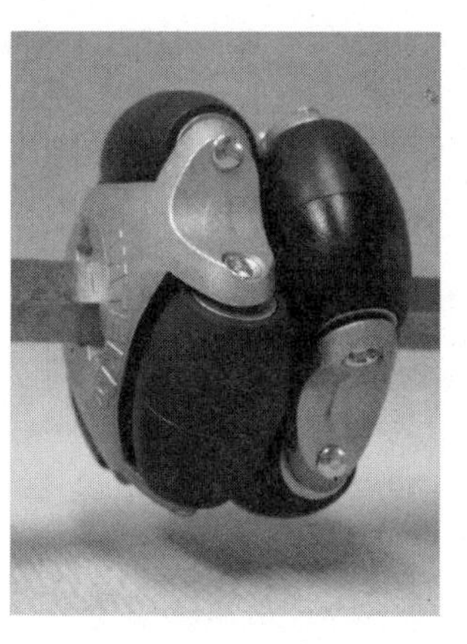

Figure 8-22 Omnidirectional wheel

Figure 8-23 Spherical Wheels

Figure 8-24 Tail wheel

One popular alternative to the swivel caster is the tail wheel shown in Figure 8-24, originally meant for radio-controlled model airplanes (and therefore available at most hobby stores). The wheels are available in sizes ranging from about $^{3}/_{4}$" to over 2", to match the scale of the airplane they are used with. The wheel is meant to be used with specific mounting hardware.

One problem with tail wheels, however, is that they generally don't swivel as well as most swivel casters. You can improve the swivel action by making sure the tail wheel assembly is well lubricated, and that the collars used to keep the assembly together are not on too tight.

Successful Use of Casters

To be effective, the caster(s) on your robot must not impede the direction or speed of the machine's travel. Cheap swivel casters can catch and not swivel properly when the robot changes direction. This can cause the robot to veer off course, because the robot wants to go one way, but the caster is still pointing in another!

When purchasing casters, it is a good idea to test them for smooth swivel action. Casters with ball bearings—as opposed to a simple plastic insert—tend to provide better results. Even with ball bearings, the swivel action may be stiff or sporadic, so always test first. Casters can be tested by rolling them on the floor and making them turn in different directions. If the caster does not quickly and effortlessly reorient itself in the direction of new travel, select a different caster.

In my experience, the lighter twin wheel casters often provide superior performance. The dual wheels help to keep the robot on track. Small twin wheel casters can be more difficult to find, however, and most have push-in stem mounts rather than threaded stems or plates. Good sources for twin wheel casters include mechanical surplus mail-order houses (such as C&H Sales, Herbach and Rademan, American Science and Surplus, and others) and furniture supply houses.

For the latter, try Thomas Register at ***www.thomasregister.com***, and search for caster manufacturers and distributors. If you ask nicely enough, many of these businesses will send samples, and one or two are all most robots need anyway.

Selecting the Right Size of Caster

If the robot's drive wheels are mounted along the center line of the robot, then two casters are used. One is placed toward the front of the robot, and the other toward the back. If the drive wheels are placed at one end of the robot, then only one balancing caster need be used.

For small roving robots, the caster can be kept fairly small. The smallest commonly available caster has a wheel of about 1$^5/_8$"; smaller models are available, but are not as easy to find. The 1$^5/_8$" caster works well with robots that have 2" to 3" diameter drive wheels.

As a general rule, the larger the drive wheel is, the larger the caster. This doesn't always hold true, however, depending on whether the motors and wheels are mounted on the top of the robot's base, or below it. When mounted on top, there is less ground clearance, so the caster must be smaller. Figure 8-25 demonstrates the motor, wheel, and caster relationship.

A problem with the very small casters: Most aren't made very well. They tend to be quite cheap, and the swivel mechanism may catch and drag. As noted in the previous section, when a swivel caster doesn't swivel, the robot can veer off course.

If your robot is small and light (e.g., under a few pounds) you should also avoid casters where the swivel is stiff because of heavy grease. Otherwise, the caster may not turn readily under a lightweight robot. Heavy grease is used for casters designed for larger loads.

One Caster or Two?

A common design approach for wheeled robots is to use two casters, one on each end of the base, as shown in Figure 8-26. This design presents several problems. The most critical of which is that the robot can become trapped if the casters touch the ground, but the drive wheels do not. This can occur, for instance, when rolling over a small dip, or simply when traveling over a carpet threshold. As the leading caster rolls over the threshold, the entire robot is lifted up. If the threshold is high enough, this can cause the drive wheels to lose contact with the ground; only the front and back caster are touching terra firma.

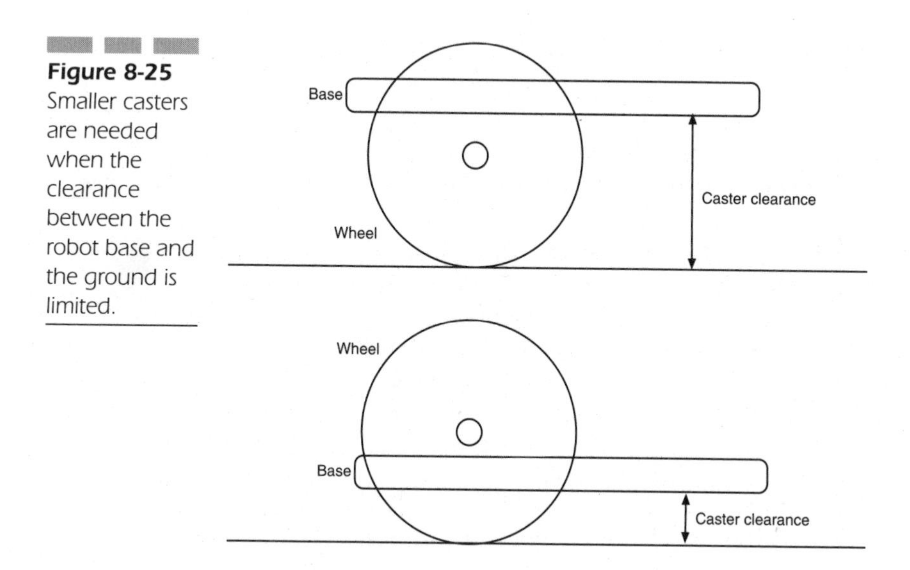

Figure 8-25 Smaller casters are needed when the clearance between the robot base and the ground is limited.

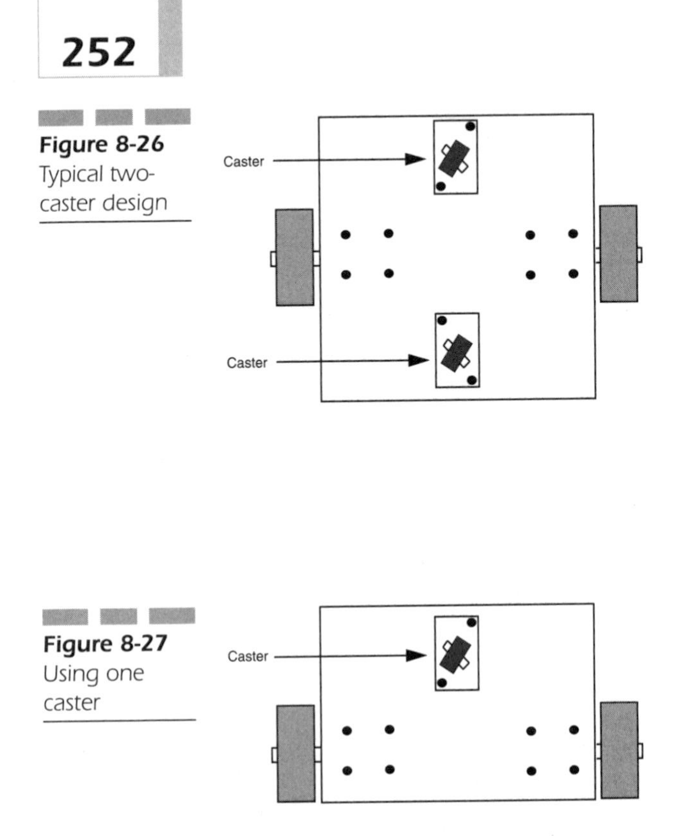

Figure 8-26 Typical two-caster design

Figure 8-27 Using one caster

There are several approaches to dealing with this problem:

- *Don't use two casters.* Instead, use one caster, and place the drive motors on the opposite end of the robot (see Figure 8-27). Such bases can then be made smaller, but note that this will not necessarily decrease the turning circle of the robot.
- *Reduce the height of the caster.* This means the robot will no longer travel on both drive wheels and both casters at one time. Instead, it will teeter-totter between the two casters, depending on whether it's going forward or backward. The caster height can be reduced by using a smaller diameter caster wheel, by increasing the drive-wheel diameter, or by using different size spacers for the motor or caster mounting.
- *Mount the casters on a spring suspension.* Compression springs can be used for this purpose. Select a spring that, under normal load, barely begins to compress from the weight of the robot. When going over a dip or obstruction, the spring will give and the drive motors will maintain contact

with the ground. A suggested approach to mounting the caster on a spring suspension is shown in Figure 8-28.

Caster Wheel Material

There are literally thousands of variations in casters. For robotics, we concentrate on a few critical details: cost, size, availability, rollability, traction, and load capacity. Cost, size, and availability are obvious, and nothing much needs to be said beyond what has already been mentioned in this chapter.

- *Rollability* is the ease with which the caster wheel turns when under load. Casters meant for a heavy-duty application may not roll well when used with a light robot. Rollability is determined by the design of the caster, as well as the caster wheel material, size and width, tread, and other factors. There is no magical formula for determining whether a given caster will operate well with a given robot.
- *Traction* is the ability of the caster wheel to grip the surface it's rolling over. This is greatly determined by both the weight on the caster and the wheel material. The harder the material (e.g., as steel or hard rubber), the less traction of the caster wheel. For high-traction applications, select a soft rubber (which tends to be gray in industrial casters), Buna rubber, or a similar material.

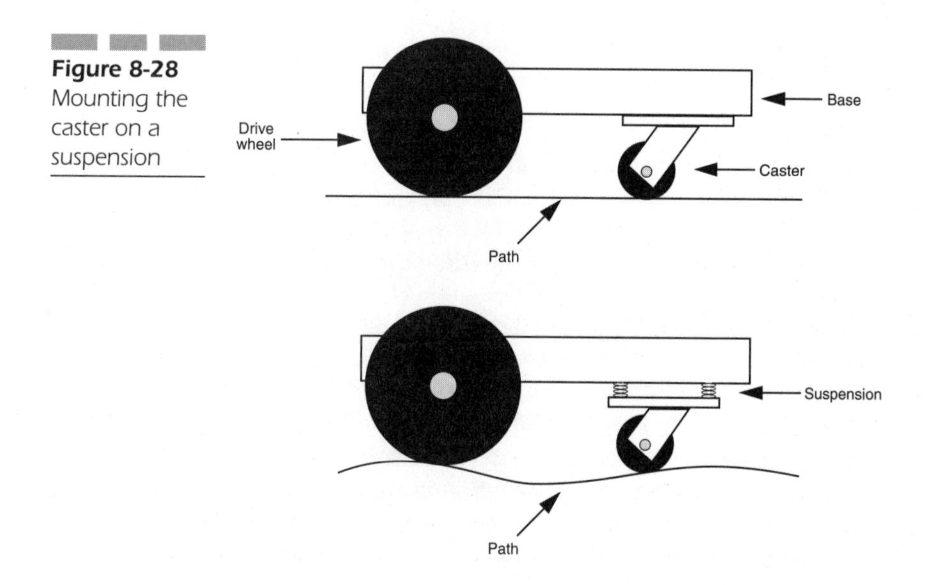

Figure 8-28 Mounting the caster on a suspension

- *Load capacity* is the weight the caster can carry. Heavier loads than that listed will reduce the functionality and life of the caster. For a robot, weight is distributed among all of its wheels and casters (this assumes the weight is equally distributed across the base of the robot). For a robot with two drive wheels and one caster, the caster is supporting one-third of the robot's weight. Your robot can weigh 150 lbs., given an average caster that supports 50 lbs.

Best Bet: Skids and Casters

For a small robot, under 4 lbs. and measuring less than about 8" in diameter, a nonrotating skid is often acceptable, if not preferable. For center-line motor mounting, use a skid in the front and back of the robot.

For larger robots, a skid may dig into soft surfaces or may snag on bumps, cables, and other obstructions. For these, use a swivel caster with a rigid metal plate, a 1⅝" to 2" diameter polyurethane wheel, and a ball thrust bearing between the plate and wheel frame. The load capacity is 10 to 50 lbs., depending on the caster.

Managing Weight

All objects on this earth have weight. Even a lightweight robot has some weight. The weight itself isn't as important as the following:

- The ratio of power to weight, that is, the ability of a robot with a given drive system to propel its own weight
- The distribution of the weight over the base of the robot

The ratio of power to weight is sometimes referred to as weight budget. Simply put, the power-to-weight ratio is the correlation between the amount of torque a robot's drive system provides, compared to the weight of that robot. Ideally, the ratio should be about even—the robot's weight is well matched to its capacity to move itself.

Weight distribution is often neglected, but equally as important. It is how the weight is placed on the base of the robot. For differentially steered robots, the most common weight distributions are shown in Figure 8-29. In all cases, the weight is centralized over the drive wheels or support casters. Avoid situations where too much weight rests over a single wheel or off-center to either the drive wheels or the caster. This may result in the robot veering significantly off course.

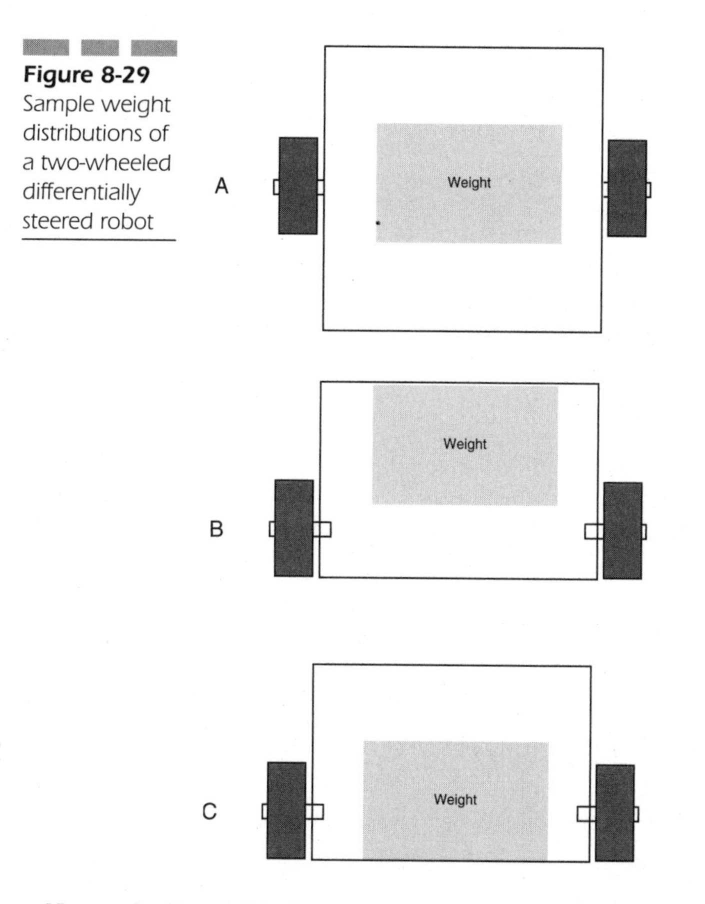

Figure 8-29 Sample weight distributions of a two-wheeled differentially steered robot

Note styles B and C in Figure 8-29. With B, the majority of the weight is over the single support caster; with C, the weight is over the drive wheels. Which is better? It depends on the amount of weight, as well as the style of caster.

- For lighter robots, position the weight over the wheels to aid in traction. If there is insufficient weight over the drive wheels, they may slip when turning.
- For heavier robots, you can try different weight-distribution arrangements to see which works best. Simply centering the weight over the base of the robot may be your best all-around choice.
- Excessive weight over swivel casters may inhibit the caster from recentering itself after a turn. If the caster fails to recenter, the robot will veer off course, as shown in Figure 8-30.

For the purposes of weight distribution, the motors and wheels, batteries, and caster(s) will represent the heaviest items. Since other aspects of the design of

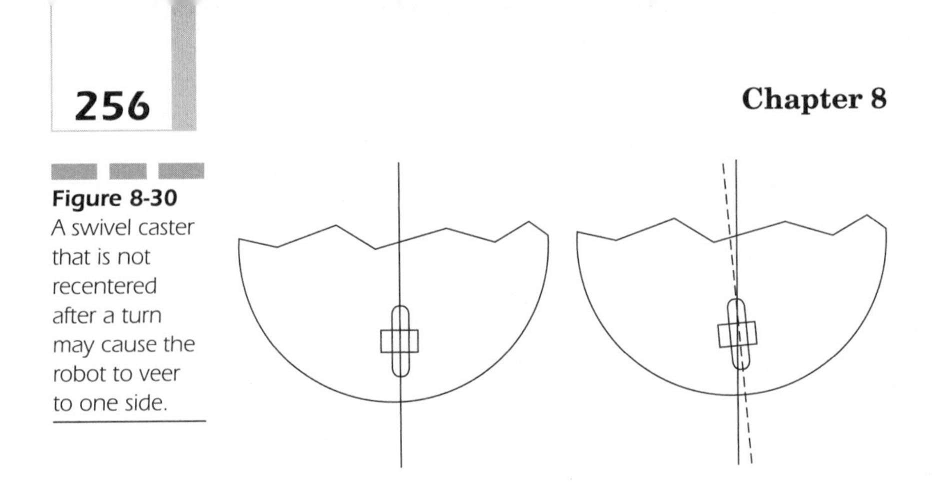

Figure 8-30
A swivel caster that is not recentered after a turn may cause the robot to veer to one side.

the robot base dictate the placement of the motors, wheels, and caster(s), this leaves the batteries as the ballast you can use to help distribute weight. When designing your robot, keep your battery-mounting options open. Place the other components (especially the lighter ones) around the batteries once you have identified the best location for them.

Commonly Used Power-Transmission Components

There are literally thousands of power-transmission components, but the following comprise the most commonly used, and the most critical.

Figure 8-31

Gears

Gears are a principle component of power transmission and are primarily used in robotics to reduce the speed, and increase the torque, of the wheel drive motors. Because of the mechanical precision required to properly mesh gears, most amateur robot builders do not construct their own gear assemblies. Gears are more fully detailed later in the chapter.

Figure 8-32

Timing Belts

Also called synchronization belts, typical timing belts for small mechanisms range from 1/8" to 5/18" width and in sizes from a few inches in diameter to several feet. Material is usually neoprene, with metal or fiberglass reinforcement. Belts are rated by the pitch between nubs or cogs, which are located on the inside of the belt. Timing belts are used with matching timing-belt pulleys, which come with either ball-bearing shafts (used for idler wheels) or with press-on or set-screw shafts for attaching to motors and other devices.

V-belts

V-belts have a tapered V shape and are used to transfer motion and power from a motor to an output when synchronization of that motion is not critical (because the belt could slip). V-belts, which are often made with metal or fiberglass reinforced rubber, are used with V-grooved pulleys. By changing the diameter of the pulleys, it's possible to alter the speed and torque of the output shaft in relation to the drive shaft. The same physics that apply to gears and gear sizes apply to V-belt pulleys as well.

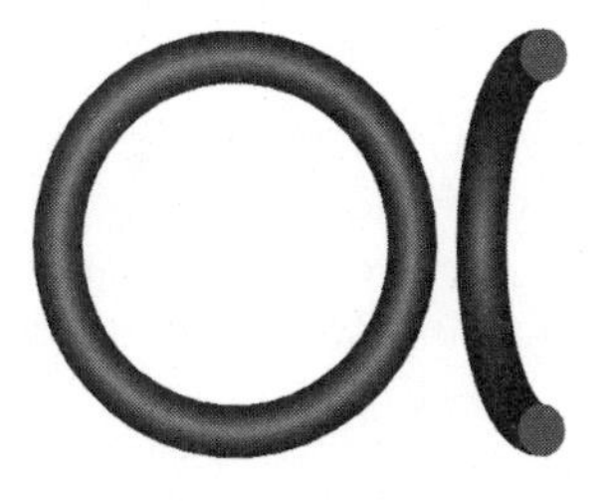

Figure 8-33

Endless Round Belts

Endless round belts are used to transfer low-torque motion. The belt looks like an overgrown O-ring and, in fact, is often manufactured in the same manner. Other endless round belts are made by fusing the ends of rounded rubber (usually neoprene). Some belt makers provide splicing kits so you can make custom belts of any length. Grooved pulleys are used with round belts and, as with V-belt pulleys, the diameter of the ground belt pulley can be altered to change the torque and speed of the output.

Ladder Chain

Ladder chain resembles a ladder and is used for fairly low-torque and slow-speed operations. Movement of a robotic arm or shoulder is a good application for ladder chain. With most chain, links can be removed and added using a pair of pliers. Special toothed sprockets, engineered to match the pitch (the distance from link to link) of the chain are used.

Figure 8-34

Roller Chain

Roller chain is exactly the same kind of chain used with bicycles, only for most small-scale machinery, the chain is smaller. Roller chain is available in miniature sizes, down to 0.1227" pitch. More common is the #25 roller chain, which has a 0.250" pitch. For reference, most bicycle chain is #50, or 0.50" pitch. Sprockets with matching pitches are used on the drive and driven components. Roller chain comes in metal or plastic; plastic chain is easier to work with and links can be added or removed. Many types of metal chains are prefabricated using hydraulic presses and require the use of master links to make a loop.

Idlers

Idlers (also called idler pulleys or idler wheels) take up slack in belt-driven and chain-driven mechanisms. The idler is placed along the length of the belt or chain and is positioned so that any slack is pulled away from the belt or chain loop. Not only does this allow more latitude in design, it also quiets the mechanism. The bores of the idlers are fitted with appropriate bearings or bushings.

Figure 8-35

Couplers

Couplers come in two styles:

- Rigid
- Flexible

Couplers are used to directly connect two shafts together, thus obviating the need for any kind of gear or belt. Rigid

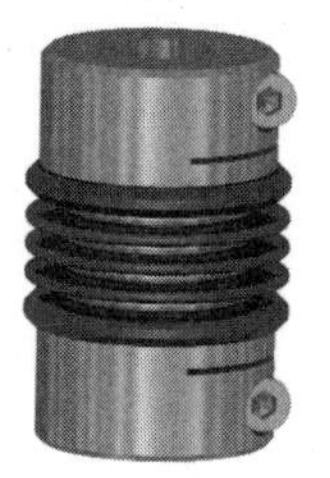

Figure 8-36

and flexible couplers are detailed more fully in their own special sections later in this chapter.

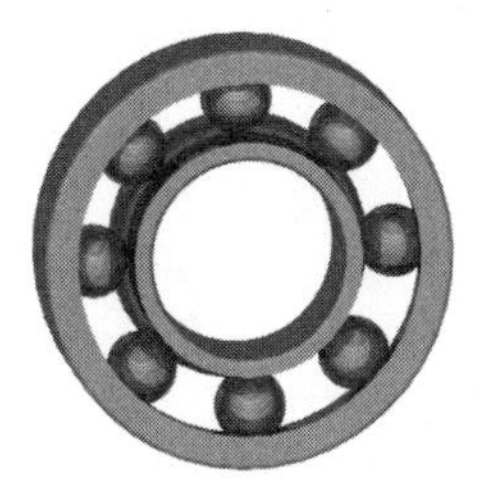

Figure 8-37

Bearings

Bearings are used to reduce the friction of a spinning component, such as a wheel or idler, around a shaft. Several bearing constructions exist, with ball bearings being the most common. The bearing is composed of two concentric rings; between each ring is a row of ball bearings. The rings—and the ball bearings—are held in place by a mechanical flange of some type. Bearings can be mounted directly to a device, which requires precision machining and a press to securely insert the bearing into place.

Another form of bearing uses narrow pieces of metal rod, called needles, which work in a similar manner.

Pillow blocks are available that allow bearings to be readily mounted on any frame or device.

Figure 8-38

Bushings

Bushings and bearings serve the same general purpose, except a bushing has no moving parts. (Note: Some people also call these *bearings* or *dry bearings,* but I prefer to use the term bushing in order to differentiate between them.) The bushing is made of metal or plastic, and is engineered to be self-lubricating.

One example is Oilite, a self-lubricated bronze metal commonly found in industrial bushings. Several kinds of plastics, including Teflon, exhibit a self-lubricating property. Bushings are used instead of bearings to reduce cost, size, and weight, and are adequate when friction between the moving parts can be kept relatively low. Bushings, and not the more-expensive bearings, are used in the output gear of the less expensive RC servos, for example.

Flexible Linkages

Flexible linkages allow mechanical power or movement to be transferred from one place to another using some form of bendable material. Examples of flexible linkages include the following:

- **Pulleys and belts** Pulleys are like wheels, and the belts ride over the wheels. Most pulleys incorporate a sleeve or rim to keep the belt in place.
- **Sprockets and chain** Sprockets are also wheels, but incorporate teeth around their circumference in order to mesh with a chain.
- **Cable** A flexible cable made of plastic or metal transfers power or movement by spinning within a protective sheath. The speedometer cable on older model cars is a good example of how these work.

Except for cable, flexible linkages can function in a similar manner to gears, including reducing or increasing speed and torque. This is accomplished by using different sprocket or pulley diameters.

One benefit of using pulleys and belts or sprockets and chain is that you needn't be as concerned with absolute alignment of the mechanical parts of your robot. When using gears, it is necessary to mount them with high precision.

Rigid and Flexible Couplers

Couplers are used to connect two drive shafts. A common application is to use a coupler to connect the drive shaft of a motor with the axle of a wheel. Connectors can be rigid or flexible. Rigid couplers are best used when the torque of the motor is low, as it would be in a small tabletop robot. Flexible couplers are advised for higher torque applications, as they are more forgiving of errors in alignment.

A rigid coupler can be made using metal or plastic tubing, selected for its inside diameter. You can purchase suitable tubing at a hobby or hardware store. Cut the tubing to length, and then drill a small hole on each end for set screws. Use a tap to thread the hole for the size of set screws you wish to use—a 4-40 setscrew is a good all-around size for most applications.

Steel tubing provides the most strength, but is harder to cut, drill, and tap. If the thickness of the tubing is sufficient, aluminum will work well for most low-torque applications. Brass and bronze should be avoided because these metals are too soft. For very low-torque jobs, plastic or even rubber tubing will work.

Select the rubber tubing so that it is slightly smaller than the motor shaft and axle you are using, and press it on for a good fit.

There are many types of commercially available rigid and flexible couplers, and the cost varies from under $1 to well over $50, depending on materials and sizes. Common flexible couplers include the helical coupler, the universal joint coupler (similar to the U-joint in the drive shafts in older cars), and the three-piece jaw coupler (more about this to follow). The couplers attach to the shafts either with a press fit, by a clamping action, by set screws, or by a keyway. Press fit and clamp are common on smaller couplers for low-torque applications; set screws and keyways are used on larger couplers.

Three-piece jaw couplers (Figure 8-39), such as those made by Lovejoy, consist of two metal or plastic pieces that fit over the shafts. These are the jaws. A third piece, the spider, fits between the jaws and acts as a flexible cushion. One advantage of three-piece couplers is that because each piece of the jaw is sold separately, you can readily mix and match shaft sizes. For example, you can purchase one jaw for a 1/4" shaft and another for a 3/8" shaft. Both jaws must have the same outside diameter.

Also note that the steering method for coaxial drive bases is exactly the same as steering a military tank: One tread stops or reverses direction while the other side keeps going. The result is that the robot turns in the direction of the stopped or reversed tread. Because of the effects of friction, differential steering is most practical with two-wheel drive systems. Additional sets of wheels can increase friction during steering.

Getting Geared Up

Of course you can always buy gears from Gears R Us. (Okay, most go by far more mundane names like Boston Gear, Small Parts, W.M. Berg, and Stock Drive.)

Figure 8-39 A three-piece jaw coupler, showing one of two metal hubs, and the center rubberized flexible cushion

You'll get what you're looking for from these sources, but it'll cost you. The average machined 1" diameter aluminum gear can cost $20 to $30.

As long as your requirements aren't too unusual, you may be able to locate the gears you want from other products and sources.

- **Toy construction sets** Don't laugh! Toys like LEGO, Erector, and Inventa come with gears you can use in your robotics projects. Most are on the large size and are made of plastic.
- **Hobby and specialty retailers** Next time you're at the hobby store, look for replacement gear sets for servos and drive motors for RC cars and airplanes. Some are plastic; others are metal (usually either aluminum or brass). Typically, you'll have to buy the whole set of replacement gears for whatever motor or servo the set is for, but in other cases you can purchase one gear at a time. Some online retailers, such as ServoCity.com and Jameco.com, sell gears specifically for hobby applications (like robots). The prices are reasonable.
- **Surplus catalogs** New gears can be expensive; surplus gears can be quite affordable. You can often find new gears, plastic or metal, for about 10 cents on the dollar, compared to the cost of the same gear new. The only problem: Selection can be limited, and it can be hard to match gear sizes and pitches even when buying gears from the same outlet.
- **Rechargeable electric screwdrivers** Inside are numerous gears, typically in a planetary configuration, used to produce their very high-speed reductions. Before raiding the screwdriver for only the gears, consider using the motor, too. The motor and gearing system of a typical electric screwdriver makes a fine robot drive system.
- **Hacked toys** Discarded and discounted toys make good gear sources. These include friction and battery-powered toy cars, dozer toys, and even some action figures. Tear the toy apart for the treasure inside. These gears tend to be small and made of plastic.
- **Old kitchen appliances** Go to thrift stores and garage sales and look for old food mixers, electric knives, and even electric can openers. Unlike toys, kitchen appliances commonly use metal gears—or at least, very strong plastic gears.

Understanding Gears

Gears are used for two purposes:

- To transfer power or motion from one mechanism to another
- To reduce or increase the speed of the motion between two linked mechanisms

The simplest gear systems use only two gears: a drive gear and a driven (or output) gear. More sophisticated gear systems, referred to as *gear trains*, *gear boxes*, or *transmissions*, may contain dozens or even hundreds of gears. Motors with attached gearboxes are said to be *gearbox motors*.

Gear Teeth

Gears are specified not only by their physical size, but also by the number of teeth around the circumference. *Spur* gears (see Figure 8-40) are most common and are used when the drive and driven shafts are parallel. *Bevel* gears have teeth on the surface of the circle rather than the edge. They are used to transmit power to perpendicular shafts. *Miter* gears serve a similar function but are designed so that no reduction takes place.

Spur, bevel, and miter gears are reversible—the gear train can be turned from either the drive or the driven end. Conversely, w*orm* and *leadscrew* gears transmit power perpendicularly and are not usually reversible. The leadscrew resembles a threaded rod.

Rack gears are like spur gears unrolled into a flat rod. They are primarily intended to transmit rotational motion to linear motion.

Gear Reduction = Torque Increase

When gears are used to reduce the output speed of a mechanism—say, a motor—the torque at the output is increased. Gears are basically a form of lever; power can be increased by changing the ratio of the lever over the fulcrum. In place of the fulcrum in a gear system is the number of teeth on each gear.

Figure 8-40 The teeth of a gear provide mechanical traction.

Gear reduction is accomplished by changing the ratio of teeth of mating gears: A two-gear system with a 100-tooth gear and a 50-tooth gear is said to have a 2:1 reduction. With such a system, output speed is reduced by 50 percent, and torque is roughly doubled.

Common Gear Specifications

Here are some common gear specifications to keep you warm at night.

- **Pitch** The size of gear teeth is expressed as pitch, which is roughly calculated by counting the number of teeth on the gear and dividing it by the diameter of the gear. Common pitches are 12 (large), 24, 32, 48, and 64 (small). Odd-size pitches exist of course, as do metric sizes.
- **Pressure angle** The degree of slope of the face of each tooth is called the pressure angle. The most common pressure angle is 20 degrees, although some gears, particularly high-quality worms and racks, have a 14½-degree pressure angle.
- **Tooth geometry** The orientation of the teeth on the gear can differ. The teeth on most spur gears are perpendicular to the edges of the gear. But the teeth can also be angled, in which case it is called a helical gear. There are a number of other unusual tooth geometries in use, including double-teeth and herringbone.

Batteries and Power Systems

Forget steam engines. Forget atomic power. Forget dilithium crystals. With few exceptions, today's robots run on battery power—the same batteries that power a flashlight, portable CD player, or cell phone. To be sure, batteries may not represent the most exciting technology you'll incorporate into your robot. But selecting the right battery for your 'bot will go a long way to enhancing the other parts that *are* more interesting.

In this chapter, we review the most common battery solutions available today. These include various types of rechargeable and nonrechargeable batteries, such as alkaline, *nickel-cadmium* (NiCad), *nickel-metal hydride* (NiMH), sealed lead acid, *lithium-ion* (Li-Ion), and polymer. We'll also cover battery packs, battery mounting, fusing, and best wiring practices.

An Overview of Power Sources

Before getting knee deep in the big muddy waters of battery selection, let's review the practical power sources available for use with mobile robots. Note the word *practical*. There are plenty of potential power sources available in the world today, but not all are suitable for the average amateur robot. Some forms of power are not suitable due to their size, safety, or cost:

- *Wind-up mechanisms* provide power using a form of tension that is slowly released. A common type is based on the idea of a clock mainspring. These use a springy metal coil as a tension spring. The spring is tensioned by winding up a key or by pulling back a pair of wheels. The coil powers a shaft or other movement as its tension is relieved. Another common windup mechanism is the rubber band, popular in toy windup airplanes. For robotics, the typical windup mechanism is confined to small toys, particularly older collectable toys.
- *Solar cells* get their power from the sun (also incandescent and many other light sources). Conventional solar cells convert the light to electricity by photovoltaic action. A disadvantage of solar cells is that power is directly related to the intensity of the light. Robots that use solar power are often equipped with a rechargeable battery or a large capacitor; both store the energy collected by the solar cell for later use.
- *Fuel cells* are gaining in popularity as an alternative energy source. Most use hydrogen in a complex chemical reaction that produces heat, as well as a flow of electricity between two electrodes. Various fuel cell designs generate different voltages—from as little as 1.5 volts (similar to a regular battery) to hundreds of volts.

- *Batteries* are by far the most common and among the least expensive methods of powering any mobile device. Batteries have been around for centuries, yet the bulk of technological breakthroughs in battery technology have taken place during the last 25 to 30 years. Batteries can be grouped into two broad categories: nonrechargeable and rechargeable. Both have their place in robotics, and cost and convenience are the primary factors dictating which to use. These issues are discussed throughout this chapter. Typical batteries, used for everything from powering modern portable CD players to amateur robots, are shown in Figure 9-1.

Chemical Makeup of Batteries

Although there are hundreds of battery compositions, only a small handful of them are regularly used in amateur robots.

Carbon-Zinc

Carbon-zinc batteries are also known as garden-variety flashlight cells, because that's the best application for them—operating a flashlight. They're a simple battery with relatively low current capacities. Although they can be rejuvenated to bring back some power, they are not rechargeable, and they end up being expensive for any high-current application, such as running a robot.

Figure 9-1 Batteries come in all shapes and sizes, but cylindrical is among the most common.

Alkaline

Alkaline batteries offer several times the current capacity of carbon-zinc batteries, and they are the most popular nonrechargeable battery used today. They cost several times more than carbon-zinc. Robotics applications tend to discharge even alkaline batteries rather quickly, so a 'bot that gets played with often will run through its fair share of cells. Good performance, but at a price.

Alkaline batteries are among the most common commercially available batteries and are sold by a variety of manufacturers. These manufacturers may use their own nomenclature to identify their batteries, or they may use any of a number of generic numbering systems, as shown in Table 9-1.

Rechargeable Alkaline

Rechargeable alkaline batteries represent one answer to the high cost of regular alkaline batteries used in high-demand applications—robotics is certainly one such application, though battery makers had products more akin to portable CD players in mind when they designed them. Rechargeable alkalines require a recharger designed specifically for the battery, and they can be recharged dozens

Table 9-1 *Alkaline nomenclatures*

Cell Size	ANSI/NEDA	Energizer	Duracell	JIS/IEC	Generic
N	910A	E90	MN9100	LR1	UM-5
AAA	24A	E92	MN2400	LR03	UM-4
AAAA	25A	E96	MN2500	—	—
AA	15A	E91	MN1500	LR6	UM-3
C	14A	E93	MN1400	LR14	UM-2
D	13A	E95	MN1300	LR20	UM-1
9-volt	1604A	522	MN1604	6LR61	PP3
6-volt	918A	521	MN918	—	—

or hundreds of times before discarding. All things considered, rechargeable alkalines are probably the best choice as direct replacements for regular alkaline cells. You can find more about why this is a factor in the section titled "Mixing and Matching Battery Voltage."

Nickel-Cadmium (NiCad)

NiCad, rechargeable batteries are an old technology and, unfortunately, one that has caused considerable poisoning of the environment—cadmium is extremely toxic. So, for the sake of limiting their lawsuits, battery makers have been weaning consumers off NiCads, instead favoring the battery formulation that follows. Although you can still get NiCads, there's little reason to, so we'll ignore them as a choice.

Nickel-metal Hydride (NiMH)

NiMH rechargeable batteries not only offer better performance than NiCads, but they don't make fish, animals, and people (as) sick when they are discarded in landfills. They are the premier choice in rechargeable batteries today, but they're not cheap. Like rechargeable alkalines, they require a recharger made for them. (Many of the latest rechargers will work with rechargeable alkalines, NiCads, and NiMH, but don't use a NiCad recharger with NiMH batteries.)

Lithium-Ion (Li-Ion)

Li-Ion cells are frequently used in the rechargeable battery packs for laptop computers and high-end camcorders. They are the Mercedes-Benz of batteries and are surprisingly lightweight for the current output they provide. However, Li-Ion cells require specialized rechargers and can be expensive. Only the most well-heeled robot experimenter can afford new Li-Ion batteries, but the rest of us can make do with dismantling a discarded or surplus laptop or cell phone battery.

Lithium

Lithium (without the -Ion) batteries are also available. These are nonrechargeable and are used for long-life applications, such as smoke detector batteries. They are also used as memory backups and are commonly available in 3-, 6-, and 9-volt cells.

Sealed Lead-Acid

Sealed lead-acid (SLA) batteries are similar in makeup to the battery in your car, except that the electrolyte is in gel form, rather than a sloshy liquid. SLAs are sealed to prevent most leaks, but in reality, the battery contains pores to allow oxygen into the cells. SLA batteries, which are rechargeable using simple circuits, are the ideal choice for very high-current demands, such as battle 'bots or very large robots.

SLA batteries tend to be quite bulky, and one of the most challenging aspects of using them is picking a size that does not overwhelm your robot. Tables 9-2 and 9-3 list representative sizes of both 6- and 12-volt SLA batteries.

Polymer

Polymer batteries are among the latest in rechargeable technology. They are used for medium- to high-current electronics applications like cellular phones. These batteries use lithium as a component, but they are not quite the same as the lithium and Li-ion cells mentioned previously. Polymer batteries can be manufactured with thicknesses as small as 1 mm wafers.

Table 9-2 Six-volt SLA battery sizes

Nominal Capacity (Amps)	Height	Width	Depth	Weight
1.0	51	42.5	54.5	0.25
2.8	134	34	64	0.59
4.0	70	47	105.5	0.85
7.0	151	34	97.5	1.29
10.0	151	50	97.5	2.0
12.0	151	50	97.5	2.1

Weight is in kilograms; size is in millimeters.

Table 9-3 Twelve-volt SLA battery sizes

Nominal Capacity (Amps)	Height	Width	Depth	Weight
0.8	96	25	61.5	0.35
1.2	97	48	55	0.59
2.0	150	20	89	0.7
3.2	134	67	64	1.23
4.0	90	70	106	1.56
7.0	151	65	98	2.8
12	151	98	97.5	3.78
17	181	75	167	5.89
18	180	76	167	6.2
24	166	175	125	8.99
38	197	164	170	13.48
65	350	166	174	22.67

Weight is in kilograms; size is in millimeters.

In Summary

Most experienced builders select from a small palette of battery types based on the size and application of their robot.

- Alkaline, NiCad, and NiMH batteries are most commonly used with small tabletop robots. When using alkalines, you may choose between the rechargeable and nonrechargeable type. If you find yourself replacing the batteries often, opt for rechargeable alkalines to save some money, or switch to NiCad or NiMH cells.
- Midsize rover robots use larger NiMH or Li-Ion cells, or the smaller capacity SLA batteries. SLA batteries are available in a wide variety of capacities, and the capacity essentially determines the size and weight of the battery. We will look closer at battery capacity later in this chapter.

- Because of the power demands, larger robots, such as those for machine combat, use SLA batteries almost exclusively. The largest robots might also use liquid electrolyte batteries originally intended for use in motorcycles, small boats, or cars. These batteries are the heaviest of the bunch.

Batteries at a Glance

As we've seen, batteries can be rechargeable or nonrechargeable, and different battery types also have different amounts of volts per cell. This is a special consideration when powering electronics and many types of motors, as providing too low a voltage may render the robot inoperable, and delivering too high a voltage (without some form of voltage regulation) may damage critical components.

In a nutshell, Table 9-4 shows the common battery types, the nominal voltage they deliver per cell (when fully charged), and other important selection criteria.

Battery Voltage and Current

The two most critical aspects of batteries are their voltage and current. The importance of *voltage* is obvious: The battery must deliver enough electrical juice to operate whatever circuit to which it's connected. A 12-volt system is best powered by a 12-volt battery. Lower voltages won't adequately power the circuit, and higher voltages may require voltage reduction or regulation, either of which entails some loss of efficiency.

Battery voltage is not absolute. The voltage of a battery may—and usually does—diminish as it is discharged. As a result, the rated voltage of the battery may vary as much as 10 to 20 percent. When fully charged, the typical 1.5-volt cell may deliver 1.65 volts; when fully discharged, the voltage may drop to 1.4 volts.

The *nominal* voltage is the normal voltage per cell for that battery. Alkaline batteries have a nominal voltage of 1.5 volts; NiCad and NiMH batteries have a nominal voltage of 1.2 volts. Only for a certain period during the battery's discharge does it actually deliver this specific voltage.

In most instances, the varying voltage of a battery does not present a problem, unless the voltage falls below a certain critical threshold. That threshold is dependent on the design of your robot, but it usually affects the electronic subsystems the most. Most electronics systems use a voltage regulator of some type, and this regulator requires some overhead (usually at least one volt, but sometimes less). As the battery voltage drops below that needed for the regulator—a predictable behavior as the battery discharges—the electronics go into a brownout mode, where it receives power, but not enough for reliable operation.

Table 9-4 Batteries at a glance

Battery	Volts/cell*	Application	Recharge†	Notes
Carbon-zinc	1.5	Low demand, flashlights	No	Cheap, but not suitable for robotics or other high-current applications.
Alkaline	1.5	Small appliance motors and electric circuits	No	Available everywhere; can get expensive when used in high-current application like robotics.
Rechargeable alkaline	1.5	Substitute for nonrechargeable variety	Yes	Good alternative to nonrechargeable alkalines.
NiCad	1.2	Medium- and high-current demand, including motors	Yes	Being phased out because of their toxicity.
NiMH	1.2	High-current demand, including motors	Yes	High capacity; still a bit pricey.
Li-Ion	3.6‡	High-current demand, including motors	Yes	Expensive, but lightweight for their current capacity.
Lithium	3	Long life, very low-current demand	No	Best used as battery backup for memory circuits.
SLA	2.0	Very high-current demand	Yes	Heavy for their size, but very high capacities available.
Polymer	3.8	Long life, medium-current demand for electronics	Yes	Cells can be made to most any size and shape; very high price; voltage varies widely over discharge.

*Nominal volts per cell for typical batteries of that group. Higher voltages can be obtained by combining cells.

†Some nonrechargeable batteries can be revitalized by zapping them with volts for a few hours. However, such batteries are not fully recharged with this method and lose their charge very quickly.

‡Li-Ion cells have different voltage characteristics, depending on manufacture; 3.6 volts per cell is common but is not considered a standard. Li-Ion batteries are almost always used in "smart" battery packs, typically delivering 7.2, 10.8, or 14.4 volts.

When selecting a battery solution for your robot, it is important to match the voltage output of the batteries with the voltage requirement of your electronics, factoring in the normal voltage drop during discharge. As an example, NiCad cells deliver a nominal 1.2 volts per cell; four cells in a typical battery pack provide 4.8 volts. At this voltage, regulation is not strictly required—most electronic systems are designed to operate at 4.5 to 5.5 volts. When fully discharged, the cells may provide only 1.1 volts per cell, or 4.4 volts for the pack. This is slightly less than the design specification for the average electronics componenets; a brownout could result.

To avoid possible brownouts:

- Use batteries with a higher per-cell voltage. For example, power the electronics from a single 9-volt battery (use appropriate voltage regulation, of course).
- Use one or more additional batteries to increase the voltage provided by the pack.
- Avoid using the batteries below their fully discharged level.
- Design for lower voltage (e.g., 3.3-volt) electronics.

If *voltage* is akin to the amount of water going through a pipe, then *current* is the pressure of that water. The higher the pressure, the more forceful the water is when it comes out. Similarly, current in a battery determines the ability of the circuit to which it is connected to do heavy work. Higher currents can illuminate bigger lamps, move bigger motors, and propel larger robots across the floor, and they can do so at higher speeds.

Because batteries cannot hold an infinite amount of energy, the current capacity of a battery is often referred to as an *energy store*, or simply as *C*, for *capacity*.

Battery capacity is rated in *amp-hours*, or roughly the amount of amperage (a measure of current) that can be delivered by the battery in a hypothetical one-hour period. In actuality, the amp-hour rating is an idealized specification: It's really determined by discharging the battery over a 5- to 20-hour period, as shown in Figure 9-2.

Few batteries can actually deliver their rated amp-hour currents throughout that stated hour. The reason? As a battery discharges, it produces heat. Heat is not only destructive to batteries (and therefore heat production may be limited by the design of the battery), but it alters the electrical characteristics of the battery. The faster the discharge, the higher the heat that is generated—if it's too fast, the battery may be destroyed, or its behavior may be unpredictable. With few exceptions, batteries are not engineered to dump all their current in a short period of time. So, manufacturers provide an idealized specification that more accurately represents the typical use of their wares.

Smaller batteries are not capable of producing high currents, and their specifications are listed in *milliamp-hours,* or mAh. There are 1,000 mAh in an amp.

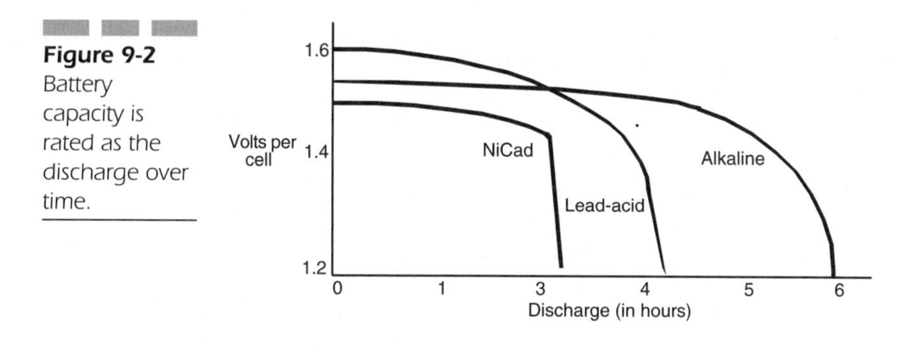

Figure 9-2 Battery capacity is rated as the discharge over time.

Therefore, a battery that delivers half an amp is listed with a capacity of 500 mAh (or less accurately as mA).

Larger batteries may be rated in mAh (as is the case with rechargeable NiCad and NiMH cells), or in amp-hours (typical of SLA batteries). Amp-hours is commonly abbreviated as *Ah* or, less accurately, as *A*.

Here are some examples:

- 2,800 mAh—2.8 Ah
- 900 mA—900 mAh
- 3.5 Ah—3,500 mAh
- 100 A—100 Ah

Very occasionally and for some applications, batteries may be rated in watts, though this is an imperfect measure. Technically, watts are calculated as voltage times current, or $V \times I$ (where V is voltage and I is current). Therefore, a battery operating at a nominal 12 volts, delivering 2 Ah of current, is rated at 24 watts.

Mixing and Matching Battery Voltage

Elsewhere we've noted that not all battery cells provide the same voltage. Alkaline cells provide 1.5 volts nominal per cell, for example. NiCads and NiMH batteries provide 1.2 volts. Most cells are used in battery packs, where they are connected in series. This acts to add the volts from each cell to provide a higher voltage from the whole pack. Therefore, four 1.5-volt batteries will yield 6 volts, but four 1.2-volt batteries will yield only 4.8 volts.

Because of this variance in cell voltage, it is not always possible to simply substitute one battery type for another in a battery pack. Circuits designed to work with a 6-volt pack may not function, or may function erratically, if used with a 4.8-volt pack. If you've been using your robot with alkaline batteries, it may not work if you substitute lower-voltage NiMH batteries.

When in doubt, check the manufacturer's datasheet (if one is available), or test the device with the lower-voltage pack. *Be sure to verify operation throughout the whole discharge period of the batteries.* The robot may work fine when the batteries are fresh, but the voltage at the battery terminals is reduced as it discharges. This means your robot could stop working after a short period of time before the batteries are fully depleted.

The Right Voltages for Your Robot

Standards are a wonderful thing. You can buy a TV at your local electronics boutique and know that it'll work when you plug it into the socket at home. Don't count on electronics for robots to be as accommodating. There is no standard for operating electronic equipment: some require 5 volts, others need 3.3 volts, and yet still others need 12, 15, 24, or 48 volts, and everything in between.

Providing the proper voltages to the various subsystems in your robot requires careful planning. Obviously, the easiest way to manage the power requirements of your robot is to choose components that operate at a single voltage—say 5 volts. That's not always possible, especially for a mechanical device like a robot, which uses a wide variety of systems.

There are three basic approaches to powering the various components in your robot. Each one is discussed here.

Single Battery, Multiple Voltages

Most of the electrical equipment in your home or office is operated from a single power supply, such as a wall current. Each piece of equipment, in turn, uses this voltage as is (as in the case of an electrically powered fan), or it converts the incoming current via a transformer or other device to the voltage required. This is the natural approach because each piece of equipment is a stand-alone unit and doesn't depend on any other to operate.

This same approach can be used in your robot. A single battery—delivering, say, 12 volts—powers different subsystems. A voltage regulator or dc-dc converter is used to provide each subsystem with the precise voltage it requires.

Although this approach sounds good in theory, in practice it can be expensive and difficult to implement properly.

- *Linear voltage regulators*, the most common variety, are cheap but relatively inefficient. In effect, they step down voltage from one level to another; the difference in voltage is dissipated as heat. The heat can be dealt with; the real problem is the unnecessary drain on the battery. It's better to conserve battery power for productive tasks, like running the robot's motors.
- *Switching voltage regulators* are more efficient—some offer efficiencies of up to 80 percent—but they are more expensive to implement, and many require additional components and design considerations. Like linear voltage regulators, switching regulators step down one voltage to provide another.
- *dc-dc converters* are self-contained voltage changers. They are the most expensive of the lot, but they require no additional components. Dc-dc converters can step down or step up voltages, and can provide negative voltages. The disadvantage of many dc-dc converters (besides cost) is that they require high input voltages in order to supply adequate current at the output. For instance, the input voltage may be on the order of 24 to 48 volts in order to provide reasonably high current at 5 or 12 volts.

Multiple Batteries, Multiple Voltages

A potential alternative to voltage regulation or conversion is to add separate battery packs to your robot. One battery pack may power the main electronics of the robot; another may power the motors. This often works well because the electronics probably need regulation, and the motors do not. The pack for the electronics can be 6 or 7.2 volts (regulated to 5 volts), and the pack for the motors can deliver 12 volts.

The trick to making this work is to tie all the ground (negative terminal) wires of the battery packs together. Each subsystem receives the proper voltage from its battery pack, but the shared grounds ensure that the various parts of your robot work together.

The exception to tied grounds is if you use opto-isolators. A typical application of opto-isolators in robotics is to control the drive motors. The electronics and the motors are on completely separate circuits, and their grounds are not tied. Rather than connect wires directly from the electronics to the motor control circuitry, the electronics instead power opto-isolators, which contain a *light-emitting diode* (LED) and a phototransistor. The link between electronics and motor control is therefore made of light, not wire.

Single Battery, Single Voltage

Depending on the subsystems of your robot, you may be able to use a single battery and single voltage for everything. For example, if your electronics do not contain any 5-volt TTL parts, you might be able to run all circuitry at 12 volts, along with the motors of your robot. Since applying excessive voltage to electronics can damage it, always check the specifications first.

The disadvantage of using a single battery for both electronics and the motor is that dc motors—especially large ones—produce a lot of electrical noise that can disrupt the operation of microcontrollers and computers. If you plan on operating your robot from a single, nonregulated battery pack, be sure to add noise suppression to the motors. One effective noise suppression technique is to solder 0.1 uF nonpolarized disc capacitors across the terminals of the motor, or from each terminal to the ground case of the motor.

Common Battery Sizes

If you've ever changed the batteries in an electronic device, you already know different sizes are available. Battery sizes have been standardized for decades (see Figure 9-3), though most consumers are familiar with only a few of the more common types: N, AAA, AA, C, A, and 9-volt. However, there are many other in-between sizes as well.

For the most part, the size of the battery directly affects its capacity—assuming the same types of batteries are compared. For example, because one C battery has roughly double the internal area of one AA battery, it stands to reason the capacity of the C battery is about twice that of the AA cell. (In actual practice, size versus capacity is more complicated than this, but it'll do for a basic comparison.)

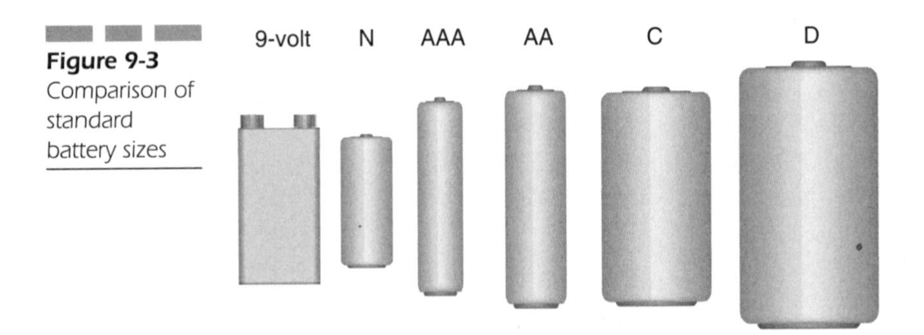

Figure 9-3 Comparison of standard battery sizes

Table 9-5 shows the typical capacity ratings for rechargeable NiCad or NiMH batteries. The capacities are by no means standard, but they are fairly typical of the battery product lines among several manufacturers. Note also that the weight can vary.

Using Ready-Made Rechargeable Battery Packs

Many of today's consumer products use rechargeable batteries. It's a fair bet that the majority of these use specially made battery packs (see Figure 9-4 for an example), rather than individual cells. Manufacturers must often specify so-called sub sizes for the cells in the pack because of size issues. It's hard to get a couple of AAs, or even AAAs, in the handset of a cordless phone, for instance.

Table 9-5 Battery capacity ratings

Cell Size	Diameter*	Height*	Weight*	Capacity in mAh
N	12.0	30.0	5	150
AAA	10.5	44.5	12	650
1/3 AA**	14.0	14.0	7	50
1/2 AA**	14.0	17.0	14	110
2/3 AA**	14.0	28.3	14	600
4/5 AA**	14.0	42.2	23	1200
AA	14.0	50.0	25	1500
A*	17.0	50.0	35	2200
1/2 C	23.0	26.0	23	2100
C	25.2	50.0	80	3500
D	32.2	60.0	150	7000
10-volt	25.7 × 17.4	48.2	45	160–200

*Diameter and height are in millimeters; weight is in grams.

**Typically used in specialty battery packs, and not available in traditional consumer packaging. They are available from battery specialty retailers as replacement cells, however.

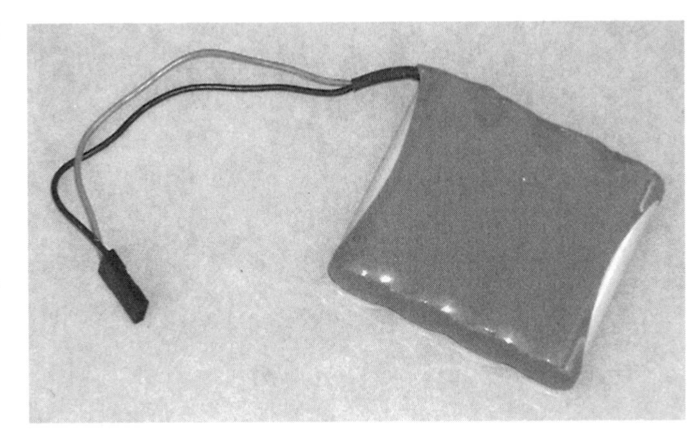

Figure 9-4 A typical battery pack for a consumer electronics device (in this case, a cordless phone)

Replacement battery packs are available for most popular brands and makes of cordless phones, cell phones, personal CD players, and other consumer electronics products. Although these replacement packs carry a premium price, there are some advantages:

- Most packs are smaller than ones you can make yourself, and are handy if space is a problem.
- Purchasing the replacement pack for a discarded device allows you to hack the recharger electronics from it. This saves you the cost and trouble of buying or building a recharger.
- If the battery pack of a discarded product is still good, you can recharge and use it, saving yourself the cost of buying new batteries.

Using Rechargeable Radio Control (RC) Battery Packs

RC applications are power hungry, and rechargeable battery packs are the norm. The battery packs are available in several voltages and current capacities. Common voltages include the following:

- 4.8 volt
- 7.2 volt
- 9.6 volt

The 7.2-volt packs are perhaps the most useful for robotics. Note the unusual fractional voltages; these are the result of the 1.2-volts-per-cell batteries used in the packs. Current capacities range from about 350 mAh to over 1,500 mAh. The higher the current capacity, the longer the battery can provide juice to your robot. Unfortunately, higher-capacity batteries also tend to be larger and heavier.

NOTE: *You should always pick the current capacity based on the estimated needs of your robot, rather than simply selecting the biggest brute of a battery that you can find.*

As mentioned earlier in this chapter, there are two general types of batteries used in rechargeable packs: NiCad or NiMH. Both can be recharged many times, but of the two, NiCad batteries are the least expensive because they've been around the longest. NiMH batteries provide for high-current capacities, with ratings of 600 mAh to 3,000 mAh (and higher).

There are other advantages to NiMH batteries. For years, users have complained about the memory effect of NiCad cells—though NiCad battery makers say this problem has long been corrected. The memory effect is simply this: If a NiCad battery is recharged before being completely discharged, it may remember this shortened current capacity. The next time the battery is used, it may not last as long as it should before needing a recharge.

Both NiCad and NiMH battery packs require rechargers specially designed for them. The better battery rechargers work with a variety of pack voltages.

Using Battery Holders

Perhaps the most convenient method of using batteries with your robot is with a battery holder. The holder provides a convenient housing for the batteries and allows for easy removal and installation. Electrical contacts in the holder form the proper series connection of each cell; the voltage at the holder terminals is the sum of all the cells.

Single-Cell and Multicell Holders

Holders are available for all the common battery sizes, and even some of the not-so-common ones. (However, if you plan on using the fractional AA size, or one of the sub sizes, you are better off building a battery pack, as detailed later in this chapter.) The number of cells the holder will accommodate varies from one to six, and in some cases eight. Holders for two and four cells are among the most common.

Battery holders come in either plastic or metal. The plastic holders are a tad more bulky, but they are lighter, and often easier to mount. Metal holders are useful in heavy-duty applications, but they cost more.

As battery holders must often conform to the shape of the object in which they are used, it shouldn't be a surprise that they are available in a variety of cell layouts—for example, a four-cell holder may orient the cells all in a row. Or it might pack them side by side. Still another holder may orient two cells side by side, both front and back. See Figure 9-5 for some examples of single-side battery holder layouts.

For the most part, there is no best design for battery holders; you need only to choose the cell layout that best suits the size and shape of your robot. An exception to this rule is when mounting the holder to your robot. It is easier to mount a one-sided holder (batteries on the front side only) directly to your robot. A holder with batteries placed on the front and back poses additional mounting difficulties.

Mounting Battery Holders to Your Robot

Single-cell and multicell battery holders can be readily secured to your robot using any of three primary methods:

- *Fasteners* provide a solid mounting. Most holders have two or more mounting holes for use with machine screw hardware. The holes tend to be

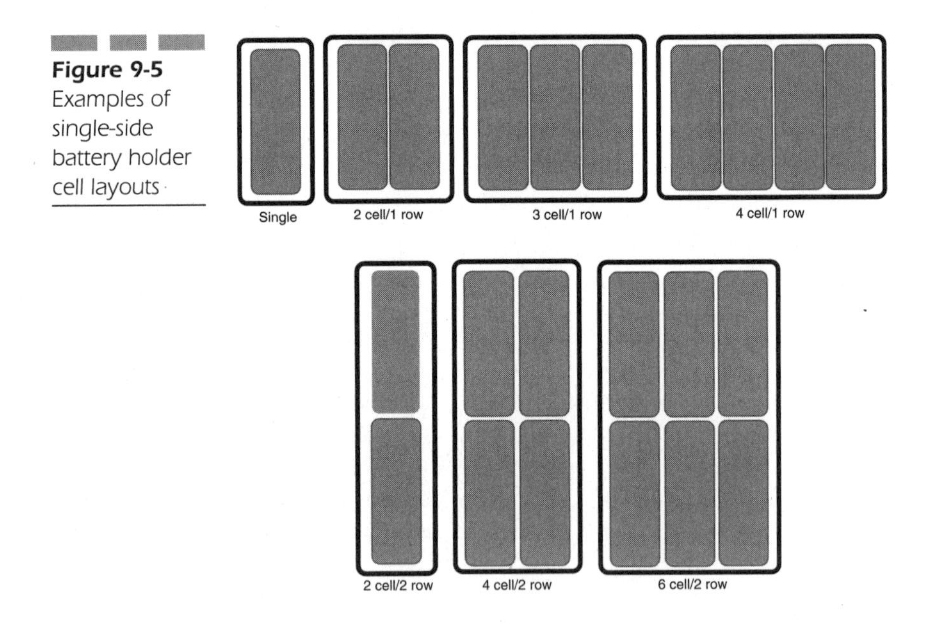

Figure 9-5 Examples of single-side battery holder cell layouts

small, so you need to either use 2-56 hardware, or drill out the holes for larger screws. Of course, you will need to drill matching holes in your robot to mount the holder. When selecting hardware, opt for flat-headed screws, and insert the screws through the holder first. This way the screw hardware will not interfere with the batteries. Fasteners are time consuming to remove, so reserve this method when you need extra holding strength and permanency, and when you can easily reach the holder to replace the batteries.

- *Velcro, Dual Lock, or similar hook and loop* allows you to mount the holder in a semipermanent fashion. The two parts of the hook and loop can be readily separated, so you need to remove the holder from the robot. This is sometimes necessary if the holder is mounted deep within the recesses of the 'bot, and you need to change the batteries. Simply remove the holder, exchange the batteries, and then press the holder back into place.
- *Double-sided foam tape* provides a quick and easy method of mounting a battery holder (see Figure 9-6). Like hook and loop, the mounting is semipermanent; even the strongest foam tape can often be dislodged with some effort. However, reserve this method for when you don't need to remove the holder from its place in your robot often. Once the foam tape has been separated, it may not stick as well when reapplied. You should use a new piece of tape.

Mount the holder in a location that allows for ready access to the cells. This allows you to replace the cells when needed. As necessary, consider the underside

Figure 9-6 Double-sided foam tape is a convenient, easy, and cost-effective way to mount a battery holder to your 'bot.

of the robot's surfaces if there isn't enough space on the top. For example, the typical two-deck robot might actually have four mounting surfaces, if you count the top and bottom sides of both decks. Most holders secure the cells with a tight fit, so you can mount the holder upside down, and the batteries will not fall out. Of course, test this first to be sure. If the batteries do have a tendency to fall out, you may need to strap them in with a one-piece Velcro strap (like the kind used in cheap wristwatches).

Snaps and Clips for 9-Volt Batteries

For 9-volt batteries, you can use a polarized battery snap. The snap is composed of male and female terminals, which mate with the corresponding terminals on the battery. As needed, you can use a separate battery clip that can be secured to the base or frame of your robot. The clip physically holds the battery in place. Clips are available for either a front or side mount; select the mount based on the preferred orientation of the battery. In either case, the clip can be secured to the robot using double-sided foam tape, Velcro or another hook-and-loop option, or hardware. I prefer the hardware approach when a permanent installation is required and I want to make sure the battery stays put.

Comparing Plastic Battery Holders

Following are several tables (Tables 9-6 through 9-10) that compare the average dimensions of a variety of battery holders and layouts in different cell sizes. You should use these dimensions as a rough design guide only, as holders from different manufacturers are not always the same sizes. Additionally, all-plastic holders tend to be slightly larger than their equivalent metal holders. In the tables, *length* (L), *width* (W), and *height* (H) are all in mm.

Table 9-6 N cells

# Cells	Orientation	L	W	H
1-cell	-	35	13	12

Table 9-7 AAA cells

# Cells	Orientation	L	W	H
1-cell	-	50	13	12
2-cell	Side-by-side	53	25	13
3-cell	Side-by-side	54	38	15

Table 9-8 AA cells

# Cells	Orientation	L	W	H
1-cell	-	57	16	13
2-cell	Side-by-side	58	32	16
2-cell	Back-to-back	57	26	17
4-cell	Side-by-side	58	63	16
4-cell	End-to-end + Back-to-back	109	26	17
4-cell	Side-by-side + Back-to-back	56	31	28
6-cell	Side-by-side	58	91	16
6-cell	Side-by-side + Back-to-back	58	45	28
8-cell	Side-by-side	59	121	17
8-cell	Side-by-side + Back-to-back	58	63	28

Table 9-9 C cells

# Cells	Orientation	L	W	H
1-cell	-	62	30	25
2-cell	Side-by-side	62	56	23
4-cell	Side-by-side	59	106	25
4-cell	End-to-end + Side-by-side	108	54	23
6-cell	End-to-end + Side-by-side	159	57	25

Table 9-10 D cells

# Cells	Orientation	L	W	H
1-cell	-	69	36	28
2-cell	Side-by-side	71	72	31
4-cell	Side-by-side	73	141	31
4-cell	End-to-end + Side-by-side	133	74	31

Building and Using All-in-One Battery Packs

As we've seen, one method of providing battery power to your robot is using single cells mounted in battery holders. This is a convenient approach—depending on the holder, you can remove the batteries for exchange or recharge, or recharge all the batteries while still in the holder.

A disadvantage to batteries in holders is that you must settle on a standard size. This is not always possible or practical. The shape and layout of your robot may make it difficult for you to use a standardized holder. In these instances, you're better off using an all-in-one battery pack, either purchased ready-made, or built by you. Because of the assembled nature of the pack, rechargeable batteries are used almost exclusively. The batteries are typically soldered together, either terminal to terminal, or via wiring.

Ready-made battery packs consisting of 3 to 10 (and more) batteries are routinely available at most any hobby store that specializes in RC airplanes, cars, and other vehicles. The packs use either NiCad or NiMH cells, and are rated by capacity, in mA. A common constituent battery size is the full AA or fractional AA (e.g., 2/3 AA) cell. A plastic sheath that is tolerant of both heat and RC engine fuel protects the cells.

Purchasing a ready-made pack is the easiest approach, and there are plenty of variations to choose from. To complement the pack you'll need a compatible battery charger that is designed not only for the total voltage of all the cells, but the type of cells, and their Ah capacity. Rechargers are available for either NiCad or NiMH (some are switchable), with common voltage ratings shown in Table 9-11.

Table 9-11 Common voltage ratings for rechargeable cells

Voltage	Number of Cells
4.8	4
7.2	6
9.6	8
14.4	10

Of course, in-between voltages are also possible, using an odd number of cells. Seven cells of either NiCad or NiMH batteries are 8.4 volts, for example.

You can also make your own packs, using any number of cells, in any available size, and in any orientation. You may assemble your custom pack with three cells side by side, for instance, and another three end to end.

Soldering cells together is the most common approach to building custom battery packs. For ease of use, get cells with solder terminals already on them; it can be hard to solder onto the + and – terminals of regular battery cells. When wire is needed, match the gauge of the wire with the current demand from the battery. Most mobile robots do not put a high stress on their batteries, so a 14 or 16 gauge stranded (not solid) wire is usually sufficient. When current requirements are higher, use a larger gauge. When you speak of a larger guage, remember, you are speaking of a lower number and a larger wire.

Use a polarized connector, also of a suitable gauge, for the positive and negative terminal ends of the batteries. Don't leave the wires bare, as this increases the chance of having a short circuit. Shorting freshly charged batteries can cause fire or burns, and can permanently damage the battery. There will be more on battery connectors later in the chapter.

The jury is out on how to bundle the batteries together. Some RC enthusiasts don't like using large heat-shrink tubing to form the protective sheath, citing the heat used to melt the tubing is detrimental to the batteries. The better RC stores sell a product invariably known as battery shrink wrap, a special thick version of heat-shrink tubing especially made for the job. When properly applied, almost no heat reaches the battery cells.

Another method is to wrap the batteries in any tape that is not conductive but will hold the cells reasonably well. Clear or brown packing tape will work. You can also assemble the cells and hold them together with small dabs of hot-melt glue. The heat of the glue is not enough to cause damage to the cells.

NOTE: When constructing your own battery packs, be sure that you have (or can get) the appropriate charger for the number and capacity of cells you are using. Obviously, it doesn't help to build a nine-cell pack if your battery charger outputs 9.6 volts only—you'd be undervolting the pack, and it will never fully recharge. The better battery chargers provide adjustable output voltages; some even automatically detect the requirements of the cell and deliver the proper voltage to recharge the batteries. These are more expensive but very versatile.

Best Battery Placement Practices

The battery in your car is intentionally mounted for easy access. Car makers know that sooner or later, you'll either need to replace the battery, apply a quick charge, or get a jump from the AAA tow truck. The same logic applies to robots. When possible, the battery holder or pack for your robot should be located where it affords quick and reliable access. This applies even if the pack is rechargeable because rechargeable batteries need to be removed occasionally for replacement, repair, or inspection.

As mentioned earlier in the chapter, given the typical mobile robot design, an ideal location for the battery pack or holder is on the underside of the robot. This assumes there is enough ground clearance between the robot and the floor. Mounting on the bottom allows for quick access to the cells, either for replacement or recharging. Simply turn over the robot.

When slinging the batteries on the undercarriage of the robot, be sure the holder or pack is securely mounted. This can be done with fasteners, double-sided foam tape, or hook-and-loop fastening. Use a temporary fastening technique if the holder or pack must be periodically removed. When using a battery holder, be sure the contact springs provide adequate pressure to keep the cells in place. If the cells are loose, strap them in using a tie-wrap, or tape them in place.

If the underside of the robot is not available for battery mounting, select a position on the robot that allows for reasonably quick access. Avoid any location that requires you to largely dismantle the robot just to access the batteries. For robots that use multiple decks or levels, place the batteries on the first level—this helps to lower the center of gravity, and generally provides a more stable base. However, be sure the top level(s) can be readily removed should you need to access the batteries. Avoid any construction or wiring that makes the batteries hard to reach.

Best Wiring Practices

Following are several easy steps to maximizing the power wiring in your robot:

- On batteries with more than 1,500 to 1,800 mAh capacity, insert a small inline fuse in the + terminal line. The fuse should be rated for approximately 30 to 50 percent more than the highest expected current draw from the robot.
- Whenever possible, use polarized (combined male-female) connectors for your battery holders or battery packs. Naturally, the connector on the holder or pack should match the battery line to your robot. Being polarized, there is less chance of connecting the battery backwards. Polarized connectors are available at hobby stores and auto supply stores.
- The ideal polarized connector fully insulates both the positive and negative sides of the power line. In this way, the battery will not short out when the connector is unplugged. This design is not always possible or practical, as it makes the connector physically large. The second best design uses an exposed negative terminal. There is less chance of a short if this terminal touches the bare metal chassis of the robot.
- Use two-color wiring to denote the polarity of the terminal. The most common color coding is red for the + terminal and black for the – terminal. Of course, you can use any wire color you wish, but strive to be consistent.
- Stranded wire is always better than solid wire. Stranded wire is more flexible and is less prone to break over time. The strands of the wire also improve the current-carrying abilities of the wire.
- Gently twist power wires to make a braid. This neatens the wiring and helps reduce the effects of radio-frequency interference.
- Use tie wraps to keep wiring together, and cable clamps to secure power leads to the body of your robot. This helps to protect the wiring from becoming snagged on objects as the robot travels across the room.

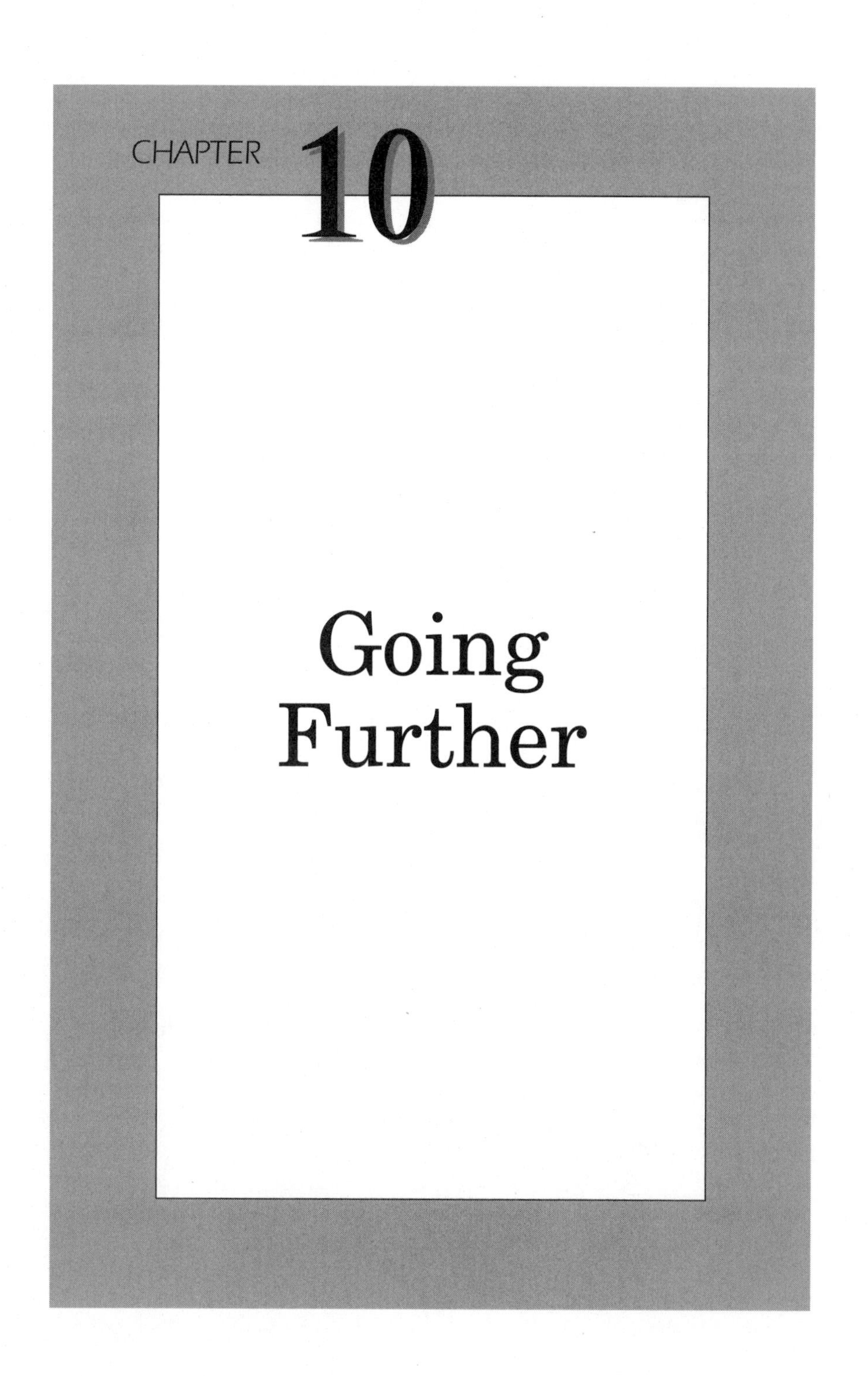

CHAPTER 10

Going Further

By now you should not be surprised that robot building comprises dozens of specialty skills and procedures. We've talked about the critical construction techniques throughout this book, yet there are more. No single book can cover every aspect of the robot craft, but that shouldn't put a damper on your enthusiasm for extending your learning. Pay frequent visits to your local library and check out all the books you can find on subjects such as metalworking and woodworking, welding, shop techniques, fabrication processes, materials, engineering, chemistry, electronics, physics . . . you get the idea.

This chapter is intended to whet your appetite for learning more about the finer aspects of robot construction. Here you'll find primer information on metal welding, brazing, soldering, plastic welding, casting with plastics and metals, using decals and transfer films, and even dressing up your robot with LEDs and electroluminescent wire.

Joining Metals by Welding, Brazing, and Soldering

As we've seen in previous chapters, metal can be joined using glue and fasteners (which include nuts, bolts, sheet metal screws, and rivets). For heavy-duty constructions, welding offers the most strength, particularly over long periods of time. Depending on the materials, and the application, it may also be practical to braze or solder the metal pieces together.

Welding is the process of fusing two or more pieces together by melting the base metal. A filler metal, usually the same as the base metal, is added to build up the weld in order to make it stronger. Given a proper weld, the join can be as strong, if not stronger, than the surrounding metal pieces.

- *Welding* metal requires very high heats, and ordinary flame torches will not do the job. A common all-purpose welding rig for mild steel is *the oxy-acetylene* torch. This torch uses acetylene as the flammable gas and oxygen as the oxidizer. Together, these gasses produce up to 5,600 degrees Fahrenheit. Oxy-acetylene torches can be used for both cutting through thinner pieces of steel and for welding. A specialty cutting torch is best used when cutting through steel, as its flame is hotter. Although oxy-acetylene welders can also be used with non-ferrous metals, the flame is generally too hot, and other types of torches are preferred.
- *Arc welders* use high-current electricity to produce a pinpoint high-temperate spot on the metal. The metal is instantaneously melted by the arc. Many arc welders require 220 volts or higher, and the larger units—used by professional welders—run off their own generators. However, there are some arc welders designed for 110-volt use. You are limited to welding fairly thin sheets of metal—down to about 12 or 14 gauge.

- *Metal inert gas (MIG)* welders are used with steel and several non-ferrous metals. MIG welders are an update of the classical arc welder, and both use high currents of electricity to join pieces of metal together. The MIG welder adds an inert gas, typically argon, to prevent oxidation during the welding process. Most MIG welding rigs use a motorized wire feed where only the right amount of filler rod is automatically pushed through the torch head. The ease of use of the MIG welder has made it a popular choice among do-it-yourselfers. The argon (or other inert gas) is readily available from industrial gas outlets. There are some 110-volt MIG welders available.
- *Tungsten inert gas (TIG)* welders are used to join aluminum, steel, and stainless steel, which are problematic metals for most of the other welding rigs. Like a MIG welder, a TIG welder uses an electric arc to produce the high temperatures for welding. The inert gas is often a mixture of helium and argon. Successful use of a TIG welder, most of which are engineered for operation at 220 volts or higher, is a matter of practice, especially when welding aluminum or stainless steel. TIG also goes by the name Heliarc.

All welders require that you wear goggles because the light of the flame (or arc) is exceedingly bright. Arc-type welders (standard, MIG, TIG, etc.) require a darker lens. For all arc-type welders, you should use a full face mask and long-sleeve shirt; the arc produces *ultraviolet* (UV) radiation, and you can get serious sunburn, even if you weld at night! Of course, all welding should be performed with appropriate welding gloves.

Brazing and soldering involve temperatures far lower than welding involves. In fact, for both brazing and soldering, the base metal being joined is not melted. Rather, the joint is created using a soft metal that melts at a relatively low temperature. The American Welding Society defines brazing as working with temperatures above 840° Fahrenheit, soldering at temperatures below 840° Fahrenheit.

Many brazing metals are silver alloys (some have copper and phosphorous added) and are commonly used to join copper plumbing pipe. Many solder metals contain lead, and these alloys melt at a lower temperature. Brazing and soldering can be accomplished using simple gas torches. In all cases, the torch is used to heat the metal to be joined, not merely the brazing or soldering metal.

- Propane torches produce a flame that is, at its hottest point, about 3,200° Fahrenheit. These are commonly referred to as air/gas torches, because they rely on the surrounding air for oxygen (air is about 30 percent oxygen; the bulk is nitrogen).
- Enhanced-gas torches, such as those with MAPP gas (a trademark), burn at higher temperatures—MAPP-gas torches burn at around 3,700° Fahrenheit, for example. Enhanced-gas torches should always be used with an approved torch head because of the increased temperatures.

- Oxygen/gas torches use separate tanks of oxygen and propane or other flammable gas. In most cases, the separate oxygen feed increases the flame temperature some 50 percent over the flammable gas used alone. A special torch head is required to use oxygen/gas mixtures.

Table 10-1 provides more information about a few types of torches.

Prior to soldering and brazing, any dirt, oil, or oxidized matter (rust or corrosion) must be removed. This is most readily done with isopropyl alcohol, followed by scrubbing with a stiff wire brush. The brush also acts to scratch up or roughen the surface and increase the surface area. The greater the surface area of the joint, the better the soldering or brazing will hold.

In almost all cases, brazing and soldering requires that flux, a waxy substance, be brushed over the surfaces to be joined. Flux is a wetting agent for the brazing or soldering material, and it allows the soldering material to flow when the base metal is heated to the correct temperature. Without flux, the brazing or soldering metal will simply clump to the base metal, making a very poor joint. Be sure to match the flux with the brazing or soldering material you are using.

Welding Aluminum

Aluminum can be welded using a MIG or TIG torch, or by the skillful use of an oxy-acetylene or arc welder. Most weekend robot builders will instead use a brazing method to join aluminum. Brazing is acceptable if high strength is not required (for example, you will want a full welding job if you plan on entering your robot in a BattleBots-style combat event).

Aluminum brazing involves the use of a hot bond alloy that melts at about 850° Fahrenheit (aluminum starts to melt at about 1,200 to 1,400° Fahrenheit, depending on the alloy). The bonding alloy joins the aluminum by surface adhesion (like a glue), not by welding. A regular propane or MAPP gas torch is sufficient for brazing aluminum.

Table 10-1 Torch types

Gas	Flame Temperature with O^2 (°F)	Cylinder Oxygen-to-Fuel Ratio	Pinpoint Heat Concentration
Acetylene	6,000	1:1	Very high
MAPP	5,300	2.5:1	Fairly high
Propane	4,800	4:1	Low

Welding Gas Safety

Acetylene, propane, and MAPP gas are highly flammable and quite dangerous when used carelessly. For obvious reasons, do not smoke when working with gas torches, and keep them away from any source of open flames. It's also a very good idea to store them separately from any flammable materials, especially cloth and motor oil.

Purchase welding gas only as you need it. Do not store it. If you've rented your oxy-acetylene tanks, and you will not need them for a while, return the cylinders to the industrial gas outlet.

Welding gas has a characteristic smell, but in low—but still dangerous—concentrations a gas leak may go unnoticed. If the concentration ignites, it could cause a serious fire or even an explosion. Keep the following in mind:

- Acetylene is lighter than air, so it rises. Avoid using an acetylene torch indoors where any leaking gas may collect near the ceiling.
- Propane and MAPP gas are heavier than air, so they sink. Any gas concentration will be near the floor, which also is where most pilot flames (from water heaters, etc.) are located.

In both cases, adequate ventilation is required to disperse any gas leakage.

Welding Plastics

As with metal, plastic can be welded rather than fastened using screws, or glued using adhesives. Plastic welding is common in cars and boats, for both original construction and repair. The most common way to weld plastics is with heated pressurized air. The joint, as well as compatible plastic welding rod, is heated by the welder, much in the same way a flame from an oxy-acetylene torch heats and melts metal. A compatible plastic can be plastic welding rod of the same or similar material being welded, or it can be strips of the material itself.

Hot-air welds can be made for all joint types, but filleting is typically required to ensure enough coverage of filler from the welding rod. The joint is made weak if there is insufficient filler material.

Some hot-air welders come complete with their own air-pressure system; others require the use of an air compressor. For plastic welding purposes, air pressure and volume can be relatively low, so a low-cost compressor—even a portable 12-volt model—is usually sufficient. Almost all commercially made hot-air welders (see Figure 10-1) come with an air-pressure regulator. Most plastics are welded at pressures of 8 to 10 *pounds per square inch* (psi.)

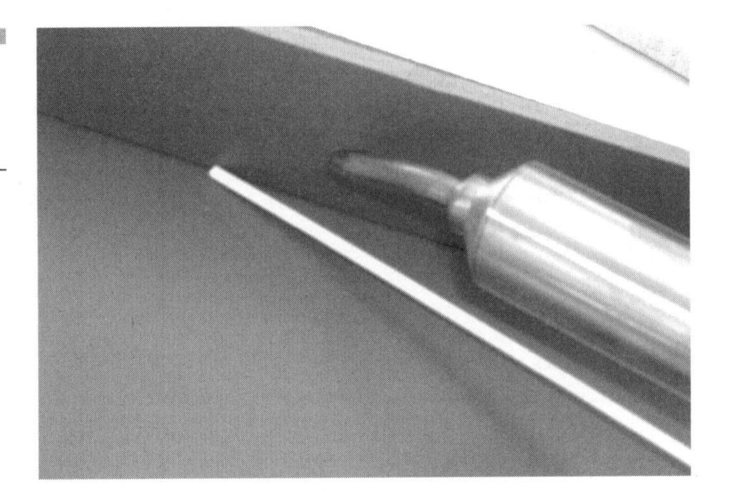

Figure 10-1 A hot-air welder for plastics

The best hot-air welders come with their own adjustable temperature control. These units are recommended so that you can control the temperature to best match the plastic you are welding. For example, PVC melts at far lower temperatures than acrylic or polycarbonate. If the temperature of the air is too high, the plastic is burned. This is dangerous because melting plastic can cause serous burns. It also produces poisonous gasses.

A method that produces weaker joints, but is still useful in some instances, is a heating mirror. The plastic is pressed against a hot slab of metal, usually aluminum. The ends soften. Before the plastic begins to reflow, the ends are removed from the heat mirror and quickly joined together. As the plastic cools, it congeals at the joint.

Other welding techniques include friction, impulse, and ultrasonic. These require specialty tools, which, for the most part, are beyond the reach of the average robot builder. However, if you have access to one of these machines at work or at school, you're sure to find a good use for it on your next 'bot project.

Casting in Plastic

Casting is a method of creating parts from a master mold. Casting can save you time and money, and if done right, it can produce a superior-looking product. Let's say you need to produce 12 sets of servo mounts for your 6-legged hexapod robot. You'll need two separate pieces for each servo, making a total of 24 mounts. You could cut and drill them all, and maybe end up with something respectable. But another approach is to cast them using a liquid plastic. Do it right and you never have to cut or drill anything.

Or, suppose you need two #25 chain sprockets, but only have one in your parts bin. With that one sprocket—the model—you can make as many duplicates as you like. Casting can be helpful if you're on a tight budget or can't wait a week or two to order a part.

Casting an object in plastic is not difficult, but it can be time consuming. The general steps are creating the mold, and then creating the casting. But of course, there's a bit more to it than that. The following is an overview.

Create the Mold

All castings require a mold of some type. If a mold does not yet exist, you will need to make one from a model. The most common molds are made of plaster (also known as plaster of Paris), which you can purchase at any arts and craft store. Specialty plasters, such as dental plaster (e.g., Hydrocal and Die-Keen) are available from mail order outlets. The mold is produced by glopping the semi-liquid plaster over the model and letting the plaster dry.

Plaster molds are either one-piece or two-piece. A one-piece mold has only one half and is suited for objects needing only the front side to be cast because the final object is meant to be placed against a flat panel (e.g., decorative trim or some kind of flat hood ornament for your robot). A two-piece mold has a front and back and is meant to cast a stand-alone figure.

Other mold materials exist, of course, and are better suited than plaster for some tasks. Plaster molds can't have any undercuts because the plaster dries hard and stiff. The more irregularly shaped the model is, the more you will need to use a flexible mold material, such as latex. Latex molds do not hold their shape for the casting step, however, and need to be reenforced with a plaster mother mold. In all, making latex and other flexible molds takes extra time.

Create the Casting

A casting is made by pouring a liquid or semiliquid material into the mold and letting the material set up and cure. Some casting materials, such as hot melt (very much like hot glue), set up when they cool. These are the fastest to work with. Others, such as urethane resin, require a chemical change that can take several hours, even overnight.

Most liquid plastics (like urethane) for casting are composed of two parts: the main plastic resin and a catalyst. Separately, the materials will never harden to produce a finished plastic piece. When mixed, a chemical reaction causes curing, which results in hardening. The catalyst must be added to the plastic in precisely the right amounts, or the casting will be ruined. The mixture is either by weight or by volume; each manufacturer will provide an instruction sheet that you should follow to the letter.

Depending on the materials you use, you may need to apply what's known as *mold release* before pouring in the liquid casting gloop. Petroleum jelly is common for plaster molds, though there are also thinner sprays available that create less mess. It is critical that the mold release be compatible with both the mold and casting materials, or the casting may never fully cure.

Playing the Waiting Game

The single greatest error made by beginners to the mold-making and casting art is trying to rush things. It can take a day or two between start and finish to complete a casting, especially if a mold is also required. And, depending on the casting materials you use, a plaster mold will need extra time—days or even weeks—for the water in the plaster to dry out.

(However, some moisture is good, as it keeps the mold from becoming too brittle. You may actually have to add water to your molds periodically, especially if you cast with urethane resins, which draw out moisture as they cure.)

Trying to rush things will result in a poor casting that either never cures (remaining soft and sticky), comes out of the mold in pieces, or is too brittle. If you take your time, you'll find your casting results will be up there with the pros.

Safety and Respirator Masks

Materials used for mold making and casting vary from harmless dust producers to downright carcinogens. You must exercise care when working with mold-making and casting materials, and remember to wear a respirator mask, safety goggles, and, if necessary, gloves.

Plaster for molds presents little danger, except that you should never use it to make a mold of an arm, leg, or other body part (yours or someone else's). Plaster cures by producing heat, and this heat can burn skin. Urethane and fiberglass resins should always be used in a well-ventilated area, where the fumes and vapors can escape out into the rest of the world. Never try to cast plastic in a closed room, and advise others that they may need to leave for a while.

Smoking is a no-no while working with molds and casting. It's important to keep all ingredients away from an open flame. Some casting materials require heating (this is the case with hot melt); use an electric double-boiler or hot plate to heat the material. Don't use an oven or range that uses gas and produces an open flame.

Using Fiberglass Casting Tape

If you've ever broken an arm or leg, the doctor probably gave you a fiberglass cast. Rather than getting out sheets of fiberglass matting, and mixing resin and catalyst in a big jug, the doctor likely ripped open a package of presoaked fiberglass tape and started wrapping it around your body. The material cures in a matter of minutes.

The benefit of using presoaked fiberglass tape is obvious, and ease of use is at the top of the list. The same casting tape is available to mere mortals. You can use it to build robust robot bodies and even to repair ones that get damaged.

3M Scotchcast is a popular casting tape available from medical supply companies. The tape is dipped in water to activate it, can be cut with scissors (before or after dipping into water), and can be worked into the desired shape over a period of several minutes before it begins to harden. Casting gloves are recommended because the tape gets sticky, and the urethane resin is a skin irritant. The tape is available in widths from 2" to 5".

Scotchcast is made for both human and animal use. Veterinary casting supplies are often cheaper. Try the following Google.com search to locate suppliers of casting tape:

- Veterinary supplies & equipment
- Casting tape
- Orthotics supply

Selecting the Right Plastic Casting Resin

There are plenty of plastic casting resins to choose from. The more you cast your own parts, the more you'll discover which casting resins are the best suited for the work you do. (You'll also discover such important things as which ones stink less.)

Table 10-2 offers a short rundown of several popular products and manufacturers. Browse each company's Web site to discover the ancillary products that might be of use to you, such as curing additives (which slow down or speed up curing), colorants, additives to change the texture and hardness of the final casting, and much more.

Table 10-2 Casting products

Product	For More Info, See These Sites
Alumlite	***www.alumilite.com***
Armorcast	***www.armorcast.com***
CR-600	***www.micromark.com***
Fast Cast	***www.goldenwestmfg.com***
Foam latex	***www.burmanfoam.com***
Polyurethane plastics	***www.polytek.com***
POR-A-KAST	***www.synair.com***
Urethane casting resins	***www.bjbenterprises.com***
Urethane casting resins	***www.smooth-on.com***

Hot and Cold Dipping Plastics

Dipping plastics are like fondue with polymers. Perhaps the best known application for dipping plastics is applying a kind of rubberized grip to hand tools—simply dip the handle of the tool into a can of dip, pull it out after a few seconds, and let it dry. Most dip can also be applied by brushing, but the idea is the same.

Dipping plastics harden to a tough shell or a rubbery sheath, depending on the chemistry used. There are two basic types: cold and hot. Cold dip is not heated before use and cures with exposure to air. Hot dip is cured by warming it in an electric oven as per the manufacturer's directions (typically 350 to 375 degrees for 15 or 20 minutes). The plastic completes its curing process when cooled. See Figure 10-2 for an example of a dipping plastic product.

For robotics, dipping plastic can be used to add a rubberized soft coating to the legs of walking 'bots, to frame bumpers to reduce damage, to wheels to add traction, and even to wire connectors to provide extra insulation.

The following are some tips when using dipping plastics:

- Curing times for cold dipping can be reduced by gently warming the dipped item with a hair dryer.
- Apply several thin coats rather than one thick coat.

Figure 10-2 Dipping plastic is available as a dip or brush-on material.

- Hot dipping is best when the object being dipped is preheated. Obviously, this won't work if you're using plastic or wood. It works well for metal objects, such as the aluminum legs in a walking robot.

The most popular dipping material is plastisol, a *polyvinyl chloride* (PVC) compound that can be made in most any color, thickness, and hardness (when cured). In addition to being used as a heavy coating, plastisol serves as a semi-flexible mold.

Casting in Metal

A far older technique than casting in plastic is casting in metal. Metal casting is found everywhere and is used to produce both small parts and large. Cast metal may be used in your automobile, the frame of an electric motor, and even a pot for cooking dinner.

Although metal casting is commonplace in industry, it's relatively rare in amateur robotics. First, cast metal tends to be quite heavy; most amateur robots are small, and it's important to keep down the weight.

Second, like plastic casting, metal casting requires a mold. The mold must withstand the high temperatures of molten metal, so making the mold is far more involved. The typical approach is to machine a mold using a steel alloy that has a higher melting temperature than the metal that will be cast. Other mold materials include sand (which can be more readily formed into a shape), investment (lost wax), ceramic, and plaster. The softer the mold, the fewer casts it will make. Some molds last for only one cast.

Sand casting is one of the most useful techniques for producing limited-use molds. There are several variations of note:

- *Green sand* molds are made by packing loose sand around a pattern and removing the pattern to form a hollow mold.
- *Shell molds* are made by heating a resin-sand mixture in metal tools, forming top and bottom shells that are joined into a form. The benefit of the shell mold is that the resin hardens, helping the sand keep its shape.

Sand casting uses disposable molds, but because the mold-making process is relatively quick and easy, for production work, several molds can be economically produced from a single original. The original can be made out of wood, plastic, plaster, or most any other material. A variety of metals can be cast using the sand methods, including iron, bronze, and aluminum.

A relatively new technique of producing finished metal pieces uses a material generically known as *precious metal clay* (PMC). Though intended for use in making jewelry, this material is also suitable for a number of robotics applications. PMC is a sculping clay that is impregnated with a fine powder of silver or gold (sometimes other metals as well). When fired, the clay burns away, leaving just the sculpted metal object. The best results are obtained using a kiln that can generate in excess of 1,500° Fahrenheit. Such kilns are available for the hobbyist at prices starting at around $300.

One use of PMC is to produce molds for casting parts using other types of plastic or metal. You can make single- or two-piece molds in this manner. The process is ideal for parts that require a fair amount of detail. The clay can be worked with your fingers or ordinary clay-sculpting tools. It will also take the impression of a 3D object, such as a gear or sprocket.

Dressing Up Your Robot

Anyone who has ever built a robot has heard the question, "So what does it do?" It's a fair question, but in most cases, it can't be easily answered. Sometimes we build robots simply because they're fun. There's no socially significant purpose for the robot, and it won't help mow the lawn, wash the dishes, or fetch a beer.

A perfectly acceptable robot is one that merely looks good. Its job is to entertain or amuse. Such show-bots occasionally make appearances at trade shows and special events, and most are really nothing more than glorified remote-controlled vehicles operated by a human. No one asks what the robot does, because they're too busy being awed by it.

An inexpensive and easy way to make your own show-bot is to add blinky lights—the quasi-technical term movie special effects masters use to dress up props. Two readily available light sources are the *light-emitting diode* (LED), and the electroluminescent wire

Light-Emitting Diodes (LEDs)

LEDs have gone well beyond small, dim, red pinpoints of light. They're now available in all colors of the rainbow, including white, deep blue, and even UV. Brightness has been drastically improved to the point where LEDs are used as flashlights. Their beam travels for hundreds of feet.

For all their luminosity, LEDs are powered by low voltage. The 5- or 6-volt battery supply to your robot will also operate the LEDs. LEDs are current-sensitive devices and will consume as much current as the battery will deliver. Because of this, they are typically used with a current-limiting resistor. The resistor restricts the amount of current that can be delivered to the LED and prevents it from being burned out. Ohm's law is used to calculate the resistor value: $R = V/I$.

Here, R is the resistance value for the resistor, V is the voltage, and I is the current (in amps). To calculate the resistor value, you first must subtract the voltage drop that occurs inside the LED from the supply voltage. Many LEDs cause a voltage drop of about 2.2 volts, so if the supply voltage is 5 volts, the value for V is 2.8.

Note that superbright LEDs, as well as LEDs in specialty colors like blue or white, exhibit a higher voltage drop. Common values are a 3.0 to 3.5V voltage drop. The voltage drop through each LED will vary between devices.

I is the amount of current you want passing through the LED. The more current, the brighter the LED will glow. But you don't want too much current or the LED will be destroyed. If the LEDs you purchase come with a datasheet, you can look up the typical and maximum current ratings for the device:

- The *continuous forward current* will tell you the recommended current of the device at normal operating temperatures and conditions.
- The *peak forward current* will tell you the absolute maximum current of the device, at normal operating temperatures and conditions, before risking a catastrophic breakdown.
- The *forward voltage* is the voltage drop through the LED.

Most designers opt for the empirical method of trying a common value of about 25 to 35 mA, and even going as far as burning out a test LED to see how much current it can handle.

We'll assume a current of 35 mA, which is expressed in the formula as 0.035 (35 milliamps is 0.035 of an amp). Doing the math, $R = 2.8/0.035$, results in a resistor value of about 80 ohms. Choose the closest standard resistor value that is equal to or higher than this. A resistor rated at 1/8 or 1/4 of a watt is sufficient.

Multiple LEDs Don't be content to use only one LED. Use multiple LEDs, of the same or different color, mounted at various places on your robot. From a calculation standpoint, it's easiest to connect each LED to the robot power supply through its own current-limiting resistor. This also helps assure an even brightness from each of the LEDs. As long as you stay under the peak forward current specification of the LED, you can vary the value of the current-limiting resistor to change the brightness of each LED.

Using Superbright and Ultrabright LEDs The typical LED produces a fairly low amount of light—a few millicandles (a *candle* is a standard unit of light measurement; a millicandle is 1000th of a candle). Superbright and ultrabright LEDs produce 500, 1,000, 2,000, 5,000, and 10,000 millicandles and higher. Some are so bright that they can cause eye damage if you stare into their beam.

Superbright and ultrabright LEDs are particularly striking on small robots. Turn down the lights and let your 'bot roam the floor. If you have a camera with an open-shutter (also called open bulb) feature, you can take a long-exposure picture that shows the path of the robot around the room.

When selecting superbright and ultrabright LEDs, I recommend paying particular attention to the beam pattern. The brightest LEDs have a narrow beam pattern—only 10 or 15 degrees. Select a broader beam pattern if you want the LED to be visible from different viewing angles.

Using UV LEDs UV LEDs have a purple-blue cast and produce long-wave UV light (about 400 nanometers) similar to that of a black light. By themselves, UV LEDs aren't all that exciting; however, they can produce spectacular light effects when illuminating an object that fluoresces. Clothes, carpets, and other fabrics with optical brighteners will appear to glow under the light of a UV LED. Certain dyes, paints, and inks are fluorescent under UV light, giving off a rainbow of colors, depending on their chemistry.

As with superbright and ultrabright LEDs, you shouldn't get into the habit of staring straight into the beam of a UV LED. Though the UV light is long-wave light and is fairly safe (as opposed to short-wave, which can cause sunburns and blindness), you will get better results if the LED is pointed down at the ground, where it will irradiate objects it passes over.

EL Wire

Electroluminescent (EL) wire is best described as a flexible neon sign. EL wire looks a lot like small plastic tubing, but when electricity is applied to it, it glows in a rainbow of colors. Applications for EL wire include the following:

- **Line following** Rather than a black or white line on the floor, the robot can follow the glow of the EL wire (put a sheet of clear acrylic plastic over the wire so the robot travels over a smooth surface). Because EL wire is available in colors, several line-following robots can trace their own color-coded line.
- **Robot-to-robot communications** A 1- to 2-foot strand of EL wire can be wrapped around your medium-sized robot and provide an omnidirectional

semaphore signal for other robots. Each robot can be distinguished by the color or pulse pattern of its wire (think Morse code, only faster).

- **Object illumination** Select a greenish or yellowish EL wire color, and use it to provide a soft illumination of objects for a robot equipped with *cadmium sulphide* (CdS) cells. Most CdS cells exhibit a peak spectral response at about 550 nanometers, which is toward the green/yellow portion of the spectrum. Because of their heightened sensitivity to this light, even a small length of EL wire produces adequate lighting for many object detection applications.
- **Robot beacon and navigation** On a similar theme, you can place EL wire along a wall baseboard; the robot can follow the glow of the wire to navigate within a room, or even from room to room.
- **Decoration** EL wire looks cool! Wrap the wire around your robot, or use the thinner EL wire as glowing whiskers for your robopet. Let your imagination run free—there is no limit to how you can dress up your robot with EL wire.

Inside EL Wire At the center of EL wire (Figure 10-3) is a solid copper conductor. This conductor is coated with an electroluminescent phosphor. To excite the phosphor, two very fine wires—or electrodes—are wrapped around the center conductor. Over this whole arrangement is a clear plastic sheath, which protects the phosphor-coated wire and electrodes inside.

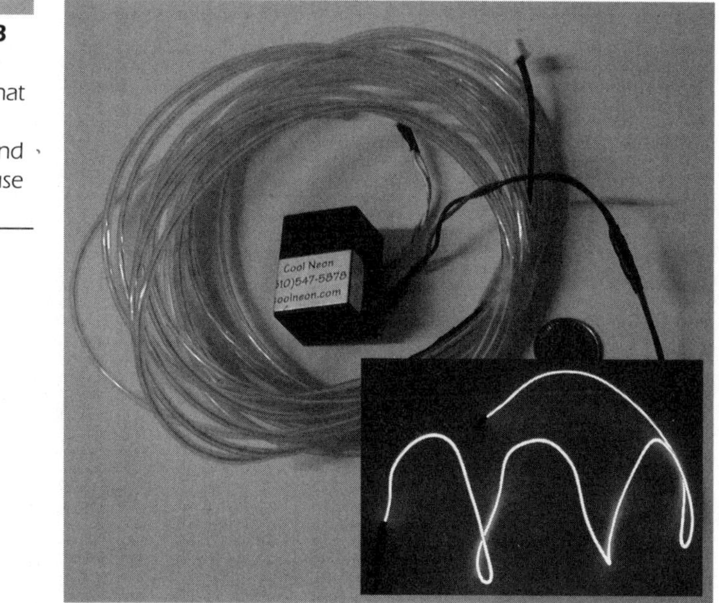

Figure 10-3 EL is flexible glow wire that has both a functional and decorative use in robotics.

Apply current to the wires, and the phosphor lights up. Colors are produced by varying the chemical makeup of the phosphor, and also by altering the tinting of the plastic protective sheath, by varying the voltage, or by varying the frequency of the current driving the wire. The end result is a brightly colored glowing wire.

Powering EL Wire EL wire is driven by a high voltage ac. But it need not be plugged into a wall outlet. Rather, the wire uses small self-contained inverters that produce the required voltage from a small dc source (usually 3 to 12 volts; AA batteries are sufficient).

Inverters are rated by their output capacity, which in turn determines the length of EL wire that can be driven. Small inverters can drive from 3 feet to 6 feet of EL wire; higher capacity units drive 20 feet to 30 feet (and more) of wire. Models designed for commercial lighting can drive several hundred feet.

Inverters are not terribly expensive—consumer models retail for $7 to $12. You'll have good results if you add more inverters to drive additional strands of EL wire, rather than try to do it all from one unit. Additionally, you can opt for an inverter that causes the EL wire to blink at specific intervals, or keeps it on continuously. Specialty inverters are available with built-in sequencers that selectively activate several strands of EL wire in turn.

Note that inverters are available at different operating frequencies—from 400 Hz to over 12,000 Hz. The brightest outputs are provided at the higher frequencies. The color of some phosphors can be altered by changing the frequency of the ac power signal. For example, the blue phosphors can be changed from green to blue by varying the frequency between 400 Hz and about 6,000 Hz. Changing the input voltage of the inverter, which in turn changes the output voltage, also alters the color of some EL wire.

Available EL Wire Colors Color choice varies by manufacturer, but most manufacturers offer the following, in diameters from 1.3 mm (called angle hair) to 5.0 mm:

- Aqua (blue/green)
- Deep red
- Green
- Indigo (deep blue)
- Lime green
- Orange
- Pink
- Purple
- Red
- White
- Yellow

The blues and greens tend to be the most vibrant colors.

One caveat when working with EL wire: Exposed (cut) ends of the wire can let in moisture, which can ruin the phosphor coating. To prevent this, always apply heat-shrink tubing over the ends of the wire. Attach the heat-shrink tubing so that it forms a closed seal.

Electroluminescent Wire Suppliers Most EL wire is manufactured by one company, ELAM (based in Israel), but it is available from a number of suppliers. The following are some that sell it mail order via the Internet:

- Cool Neon ***www.coolneon.com***
- Coolight ***www.coolight.com***
- Glowire ***www.glowire.com***
- Lightgod.com ***www.lightgod.com***
- Surelight ***www.surelight.com***
- That's Cool Wire ***www.thatscoolwire.com***
- Xenoline ***www.xenoline.com***

Color Anything with Water-Slide Decals

If you've ever assembled a plastic model kit, you know about water-slide decals: Dip the decal into warm water, wait a few seconds, and then slide the clear decal emulsion off the backing paper and onto your model.

Modern marvels being what they are, you can print your own water-slide decals using a laser or inkjet printer. The decal paper can accept black and white or color images. Although you must print a full sheet at a time (though some printers will accept smaller sizes down to about 4" by 6"), you can cut out individual decals and apply the smallest of pieces to various parts of your robot.

You can use water-slide decals to make control panels or labels and, of course, to dress up your robot with color or design. The decal will stick to most any smooth surface and should be fixed into place with a gloss or matte overspray. These chemicals are available at any hobby or art supply store and are expressly designed for use with decals.

Water-slide decal paper is but one subgroup of a larger family of printer transfer films, which also include clear and translucent "sticky-back"—used primarily for overlays—and adhesive-backed opaque vinyl, often used for bumper stickers and signs. These materials share the common trait of compatibility with color copiers and laser or inkjet printers, allowing you to prepare your own designs quickly and relatively inexpensively.

Suppliers of water-slide decal paper and similar products include the following:

- SuperCal Decals ***www.supercaldecals.com***
- Bel, Inc. ***www.beldecal.com***
- Lazertran ***www.lazertran.com***
- Papilio Supplies ***www.papilio.com***

Using Transfer Films

Similar in application to water-slide decals are transfer films. Unlike decals, however, transfer films are designed to be applied using a variety of techniques, such as pressure or heat. Thanks to a wide variety of transfer films, expressing your creativity and showing it to the world has never been easier. With only a laser printer, inkjet printer, or copier, you can transfer images to many kinds of surfaces, including fabric, wood, metal, and plastic.

Among the most common transfer films are iron-on transfers for fabric, which are used for making custom T-shirts. Elsewhere in this section you can learn about Lazertran, a water-slide decal transfer film that works with color copiers. Once printed, you can transfer the Lazertran image to a variety of hard surfaces (they also make a version for printing on fabric to make framable art).

And yet there are still others, all driven by a robust appetite for such custom-made items as mugs, mousepads, the aforementioned T-shirts, amateur sports trophies—you name it, there's probably a transfer film for it. Some, such as the Alps and Citizen Printiva, which use a dry-resin thermal transfer ink for the most vibrant colors, require a special kind of printer. Others work with ordinary inkjet and laser printers. The transfer method is usually made by heat (i.e., the toner or ink transfers from film to surface using an iron or heat press) or by immersion in liquid (usually water).

The following are some additional resources for transfer films. Be sure to check out their offerings to see what they have available.

- 1 Source Mouse Pads ***www.pilgrim-co.com***
- ACP Technologies ***www.acp.com***
- Sawgrass Systems ***www.sublimation.com***
- Screen Web ***www.screenweb.com***
- SSK Sign Supply ***www.ssky.com***

Other Google.com search phrases you might want to try include the following:

- Sublimation transfer film
- Inkjet transfer film
- Thermal transfer transfer films

APPENDIX A

Resources

Selected Robot Books

Amphibionics: Build Your Own Biologically Inspired Reptilian Robot
Karl Williams; ISBN: 007141245X

Applied Robotics
Edwin Wise; ISBN: 0790611848

Building Robot Drive Trains (Robot DNA Series)
Clark and Owings; ISBN: 0071408509

Building Robots With Lego Mindstorms
Ferrari, Ferrari, and Hempel; ISBN: 1928994679

Build Your Own Combat Robot
Miles and Carroll; ISBN: 0072194642

Build Your Own Robot!
Karl Lunt; ISBN: 1568811020

Combat Robots Complete: Everything You Need to Build, Compete, and Win
Chris Hannold; ISBN: 0071408886

Creative Projects with LEGO Mindstorms
Benjamin Erwin; ISBN: 0201708957

Dave Baum's Definitive Guide to LEGO Mindstorms
Dave Baum; ISBN: 1893115097

Five Hundred and Seven Mechanical Movements
Henry T. Brown; ISBN: 1879335638

Home Machinist's Handbook
Doug Briney; ISBN: 0830615733

Illustrated Sourcebook of Mechanical Components
Robert O. Parmley (Editor); ISBN: 0070486174

Insectronics: Build Your Own Walking Robot
Karl Williams; ISBN: 0071412417

JunkBots, Bugbots, and Bots on Wheels: Building Simple Robots With BEAM Technology
Hrynkiw, Tilden; ISBN: 0072226013

Lego Mindstorms Interfacing
Don Wilcher; ISBN: 0071402055

Mechanical Devices for the Electronics Experimenter
Britt Rorabaugh; ISBN: 0070535477

Mechanisms and Mechanical Devices Sourcebook
Sclater and Chironis; ISBN: 0071361693

Mobile Robots: Inspiration to Implementation
Jones, Flynn, and Seiger; ISBN: 1568810970

Programming Robot Controllers (Robot DNA Series)
Myke Predko; ISBN: 0071408517

The Prop Builder's Molding & Casting Handbook
Thurston James; ISBN: 1558701281

Robot Builder's Bonanza, Second Edition
Gordon McComb; ISBN: 0071362967

Robot Builder's Sourcebook: Over 2,500 Sources for Robot Parts
Gordon McComb; ISBN: 0071406859

Robot Building for Beginners
David Cook; ISBN: 1893115445

Robotic Explorations: An Introduction to Engineering Through Design
Fred Martin; ISBN: 0130895687

Robot Mechanisms and Mechanical Devices Illustrated
Paul E. Sandin; ISBN: 007141200X

Robots, Androids and Animatrons, Second Edition
John Iovine; ISBN: 0071376836

Tabletop Machining
Joe Martin; ISBN: 0966543300

Sources—Online and Mail-Order

Adhesives and Glues

Cyberbond
www.cyberbond1.com

Hot Melt City
www.hotmelts.com

IPS Corp.
www.ipscorp.com

Loctite Corp
www.loctite.com

Arts and Crafts

ArtSuppliesOnline.com
www.artsuppliesonline.com

ASW—Art Supply Warehouse
www.aswexpress.com

Dick Blick Art Materials
www.dickblick.com

Hobby Lobby
www.hobbylobby.com

McGonigal Paper & Graphics
www.mcgpaper.com

Michael's Stores, Inc.
www.michaels.com

MisterArt
www.misterart.com

Sax Arts & Crafts
www.artsupplies.com

T. N. Lawrence & Son, Ltd.
www.lawrence.co.uk

Books, Internet Resources

Amazon.com
www.amazon.com

Barnes & Noble
www.bn.com

McGraw-Hill Professional
www.books.mcgraw-hill.com (Publishers of this book, and many other fine books on robotics)

Robotbooks.com
www.robotbooks.com

Robotics Universe
www.robotoid.com (Online support site for my robotics books)

Robot Palace
www.robotpalace.com

Casting and Mold Making

Abatron, Inc.
www.abatron.com

Alumilite Corporation
www.alumilite.com

American Art and Clay Co., Inc.
www.amaco.com

Armorcast
www.armorcast.com

Bare-Metal Foil Co.
www.bare-metal.com

Castcraft
www.castcraft.com

Douglas and Sturgess, Inc.
www.artstuf.com

Eager Plastics, Inc.
www.eagerplastics.com

Freeman Manufacturing & Supply Co.
www.freemansupply.com

GoldenWest Manufacturing
www.goldenwestmfg.com

Miniature Molds.com
www.miniaturemolds.com

Perma-Flex Mold Co., Inc.
www.perma-flex.com

Schenz Theatrical Supply, Inc.
www.schenz.com

Smooth-On
www.smooth-on.com

Special Effect Supply Co.
www.fxsupply.com

Engineering Resources

Calculator City
www.1728.com

Formulas, Assorted
www.textrolinc.com/formulas.html
www.alanwire.com/tech/misc.html
www.powertorque.com/engineering_info/p_formulas.htm

Gear Formulas
www.pic-design.com/tech/gear_form/gear_form1.htm
www.tsrsoftware.com/gear-formulas.htm
www.sdp-si.com/D190/HTML/D190T123.htm

Motor Formula Application
www.joliet-equipment.com/formulas_motor_application.htm

Motor Formulas
www.elec-toolbox.com/formulas/motor/mtrform.htm

Online Conversion
www.onlineconversion.com

Torque Formulas
www.iprocessmart.com/techsmart/formulas.htm

Weight Calculator/Plastics and Metal
www.matweb.com/weight-calculator.htm

Fasteners

Aaron's General Store
www.aaronsgeneralstore.com

Allmetric Fasteners, Inc.
www.allmetric.com

Bolt Depot
www.boltdepot.com

BoltsMART
www.boltsmart.com

Fastenal Company
www.fastenal.com

Fastener-Express
www.fastener-express.com

MSC Fasteners
www.mscfasteners.com

Small Parts, Inc.
www.smallparts.com

Smith Fastener Company
www.smithfast.com

Fiberglass and Composites

Aerospace Composite Products
www.acp-composites.com

Air Dynamics
www.airdyn.com

Art's Hobby
www.arts-hobby.com

Dave Brown Products
www.dbproducts.com

Michigan Fiberglass
www.michiganfiberglass.com

Foam Materials

Advanced Plastics, Inc.
www.advanced-plastics.com

Plaster Master Industries
www.plastermaster.com

R & J Sign Supply
www.rjsign.com

Sign Foam
www.signfoam.com

Gears and Other Power Transmission Supplies

Bearing Belt Chain
www.bearing.com

Boca Bearing
www.bocabearings.com

Boston Gear
www.bostgear.com

Dura-Belt, Inc.
www.durabelt.com

Go Kart Supply
www.gokartsupply.com

igus GMBH
www.igus.de

Lovejoy, Inc.
www.lovejoy-inc.com

Manufacturer's Supply, Inc.
www.mfgsupply.com

Power Transmission.com
www.powertransmission.com

Servo City
www.servocity.com

Serv-o-Link
www.servolink.com

Small Parts, Inc.
www.smallparts.com

Stock Drive Products
www.sdp-si.com

W.M. Berg
www.wmberg.com

Hardware, General

Ace Hardware
www.acehardware.com

Aubuchon Hardware
www.aubuchon.com

B&Q
www.diy.com

CornerHardware.com
www.cornerhardware.com

Home Depot
www.homedepot.com

Lowe's Companies, Inc.
www.lowes.com

Maintenance Warehouse, A Home Depot Company
www.mwh.com

Orchard Supply Hardware
www.osh.com

LEGO

LEGO League International
www.firstlegoleague.org

LEGO Mindstorms—Home Page
www.legomindstorms.com

LEGO Mindstorms Internals
www.crynwr.com/lego-robotics/

LEGO Mindstorms SDK
mindstorms.lego.com/sdk/

LEGO Shop-at-Home
shop.lego.com

Lugnet
www.lugnet.com

Mindstorms RCX Sensor Input Page
www.plazaearth.com/usr/gasperi/lego.htm

NQC (Not Quite C)
www.baumfamily.org/nqc

Magazines

Circuit Cellar
www.circuitcellar.com

Electronic Design
www.elecdesign.com

Elektor Electronics
www.elektor-electronics.co.uk

Nuts & Volts Magazine
www.nutsvolts.com

Servo Magazine
www.servomagazine.com

Plastics Technology Online
www.plasticstechnology.com

RC Car Action
www.rccaraction.com

RC Modeler Magazine
www.rcmmagazine.com

Seattle Robotics Encoder
www.seattlerobotics.org/encoder/

Woodworking Pro
www.woodworkingpro.com

Mechanical, General

Grainger (W.W. Grainger)
www.grainger.com

McMaster-Carr Supply Company
www.mcmaster.com

MSC Industrial Direct Co., Inc.
www.mscdirect.com

Reid Tool Supply Co.
www.reidtool.com

Small Parts, Inc.
www.smallparts.com

Stock Drive Products
www.sdp-si.com

WESCO International, Inc.
www.wescodist.com

Metal Sources

Admiral Metals
www.admiralmetals.com

Airparts, Inc.
www.airpartsinc.com

All Metals Supply, Inc.
www.allmetalssupply.com

Cal Plastics and Metals
www.calplasticsand-metals.com

Industrial Metal Supply Co.
www.industrialmetal-supply.com

K&S Engineering
www.ksmetals.com

MetalMart.Com
www.metalmart.com

MetalsDepot
www.metalsdepot.com

Online Metals
www.onlinemetals.com

Sam Schwartz, Inc.
www.samschwartzinc.com

XPress Metals
www.midlandxpress-metals.com

Paper and Plastic Laminates

ASW—Art Supply Warehouse
www.aswexpress.com

GoldenWest Manufacturing
www.goldenwestmfg.com

R & J Sign Supply
www.rjsign.com

Plastics Sources

Advanced Plastics, Inc.
www.advanced-plastics.com

Advantage Distribution
www.advantage-distribution.com

ASAP Source
www.asapsource.com

Aspects, Inc.
www.aspectsinc.com

Budget Robotics
www.budgetrobotics.com

Cal Plastics and Metals
www.calplasticsand-metals.com

GoldenWest Manufacturing
www.goldenwestmfg.com

Laird Plastics
www.lairdplastics.com

Lynxmotion
www.lynxmotion.com

McMaster-Carr Supply Company
www.mcmaster.com

Multi-Craft Plastics, Inc.
www.multicraftplastics.com

Plastic Products, Inc.
www.plastic-products.com

Plastruct, Inc.
www.plastruct.com

PTG/Patios To Go
www.patiostogo.com

Public Missiles, Ltd.
www.publicmissiles.com

R & J Sign Supply
www.rjsign.com

Regal Plastics
www.regalplastics.com

Ridout Plastics
www.ecomplastics.com

Savko Plastic Pipe & Fittings, Inc.
www.savko.com

Small Parts, Inc.
www.smallparts.com

Solarbotics
www.solarbotics.com

Specialty Resources Company
www.aliendecor.com

TAP Plastics
www.tapplastics.com

Radio Control

Airtronics, Inc.
www.airtronics.net

America's Hobby Center
www.ahc1931.com

Balsa Products
www.balsapr.com

BestRC/Hobbico, Inc.
www.bestrc.com

Discount Train and Hobby
www.discount-train.com

Du-Bro Products, Inc.
www.dubro.com

eHobbies
www.ehobbies.com

Futaba
www.futaba-rc.com

Hitec/RCD
www.hitecrcd.com

Hobby Barn, The
www.hobbybarn.com

Hobbylinc.com
www.hobbylinc.com

Hobby Lobby International, Inc.
www.hobby-lobby.com

Hobby People
www.hobbypeople.net

Maxx Products International, Inc.
www.maxxprod.com

RC Yellow Pages
www.rcyellowpages.com

Servo City
www.servocity.com

Sullivan Products
www.sullivanproducts.com

Tower Hobbies
www.towerhobbies.com

Robot Kits

Acroname, Inc.
www.acroname.com

Arrick Robotics
www.robotics.com

Budget Robotics
www.budgetrobotics.com

Competition-Robotics
www.competition-robotics.com

Diversified Enterprises
www.robotalive.com

EasyBot
www.easybot.net

Evolution Robotics, Inc.
www.evolution.com

Hobbytron
www.hobbytron.net

HVW Technologies, Inc.
www.hvwtech.com

JCM Inventures
www.jcminventures.com

Johuco Ltd.
www.johuco.com

Kadtronics
www.kadtronix.com

Kelvin
www.kelvin.com

Lynmotion
www.lynxmotion.com

Mekatronix, Inc.
www.mekatronix.com

Milford Instruments Ltd
www.milinst.com

Mondo-tronics, Inc./RobotStore.com
www.robotstore.com

Mr. Robot
www.mrrobot.com

Parallax, Inc.
www.parallax.com

Robodyssey Systems
www.robodyssey.com

Robotikits Direct
www.robotikitsdirect.com

Robot Palace
www.robotpalace.com

Robot Zone
www.robotzone.com

Seattle Robotics
www.seattlerobotics.com

Solarbotics Ltd.
www.solarbotics.com

Tab Electronics Build Your Own Robot Kit
www.tabrobotkit.com

Total Robots Ltd.
www.totalrobots.com

Zagros Robotics
www.zagrosrobotics.com

Specialty Robotics Retailers

Acroname, Inc.
www.acroname.com

Budget Robotics
www.budgetrobotics.com

HVW Tech (Canada)
www.hvwtech.com

Kronos Robotics
www.kronosrobotics.com

Lynxmotion
www.lynxmotion.com

Robot Store
www.robotstore.com

Solarbotics
www.solarbotics.com

Tools, General

Campbell Hausfeld
www.chpower.com

Enco Manufacturing Co.
www.use-enco.com

Grizzly Industrial, Inc.
www.grizzly.com

Harbor Freight Tools
www.harborfreight.com

Northern Tool & Equipment Co.
www.northerntool.com

Penn State Industries
www.pennstateind.com

RB Industries, Inc.
www.rbiwoodtools.com

Rockler Woodworking and Hardware
www.rockler.com

Shopsmith, Inc.
www.shopsmith.com

Zona Tool Company
www.zonatool.com

Tools, Precision and CNC

Blue Ridge Machinery and Tools Co.
www.blueridge-machinery.com

Clisby Miniature Machines
www.clisby.com.au

Flashcut CNC
www.flashcutcnc.com

MAXNC, Inc.
www.maxnc.com

MicroKinetics Corporation
www.microkinetics.com

Micro-Mark
www.micromark.com

Minicraft
www.minicrafttools.com

Sherline Products
www.sherline.com

Super Tech & Associates
www.super-tech.com

TAIG Tools
www.taigtools.com

Techno-Isel
www.techno-isel.com

Transfer Film and Decals

Bel, Inc.
www.beldecal.com

HPS Papilio
www.papilio.com

Lazertran Limited
www.lazertran.com

SuperCal Decals
www.supercaldecals.com

Visual Communications
www.visual-color.com

Wood Sources

Constantine's Wood Center
www.constantines.com

Craft Supplies USA
www.woodturners-catalog.com

HUT Products
www.hutproducts.com

Midwest Products Co., Inc.
www.midwestproducts.com

Stockade Wood & Craft Supply
www.stockade-supply.com

Woodcraft Supply Corp.
www.woodcraft.com

APPENDIX B

Formulas and Specifications

Charts and Formulas

Table B-1 Conversion formulas

Conversion	Formula
cm of Hg	bar (atmospheres) x 76.0
cm of Hg	gm/cm^2 x 0.07356
cm of Hg	lb/in^2 x 5.1715
cm of Hg	lb/ft^2 x 0.035913
cm/sec^2	gravity x 980.665
cm^2	ft^2 x 929.0
cm^3	ft^3 x 28,317
cm^3	in^3 x 16.387
cm^3	liter x 946.358
cm^3 /sec	ft^3 /min x 472.0
°C	(°F - 32) ÷ 1.8
°F	(°C x 1.8) + 32
ft	m x 3.281
ft^2	acre x 0.000023
ft^2	m^2 x 10.764
ft^3	m^3 x 35.314
ft^3	liter x 0.03532
ft/sec	m/sec x 3.2808
ft/sec^2	gravity (sea level) x 32.174
ft/sec^2	m/sec^2 x 3.2808

Table B-1 Conversion formulas (Continued)

Conversion	Formula
ft^3 /sec	liter/min x 0.000589
ft/min	cm/sec x 1.9685
ft^3 /min	m^3 /sec x 2118.9
gravity	cm/sec^2 x 0.03108
gm	lb x 453.5924
gm/cm^3	lb/ft^3 x 0.016018
gm/cm^3	lb/in^3 x 27.680
in	cm x 0.3937
in^2	cm^2 x 0.155
in^3	cm^3 x 0.061023
in^3	liter x 61.03
in of Hg at 32F	bar (atmosphere) x 29.921
in of Hg at 32F	lb/in^2 x 2.0360
in of Hg at 32F	in of H_2O at 4C x 0.07355
in/°F	cm/°C x 0.21872
kg	lb x 0.45359
$kg\text{-}cal/m^2$	BTU/ ft^2 x 2.712
kg/mm	lb/ft x 1.488
kg/cm^2	lb/in^2 x 0.0703
kg/m^2	lb/ft^2 x 4.8824
kg/m^3	lb/ft^3 x 3.60
lb	kg x 2.2046
lb/in	gm/cm x 0.0056
lb/in^3	gm/cm^3 x 0.036127
lb/ft^3	gm/cm^3 x 62.428
lb/sec-ft	Centipoises x 0.000672
lb/hr-ft	Centipoises x 2.42

Table B-1 Conversion formulas (Continued)

Conversion	Formula
m	ft x 0.3048
mm	in x 25.40
m^3	liter x 0.001000
m^3 /kg	ft^3 /lb x 0.062428
m^3 /min	ft^3 /min x 0.02832

Table B-2 Dimension conversions, metric to inch

mm	inch
1	.039
2	.079
3	.118
4	.157
5	.197
6	.236
7	.276
8	.315
9	.354
10	.394
11	.433
12	.472
13	.512
14	.551
15	.591
16	.630
17	.669

Table B-2 Dimension conversions, metric to inch (Continued)

mm	inch
18	.709
19	.748
20	.787
21	.827
22	.866
23	.906
24	.945
25	.984
26	1.024

Multiply millimeters by .03937 to obtain inches.

Multiply inches by 25.4 to obtain millimeters.

Table B-3 Engineering unit abbreviations

Abbreviation	Meaning
ACF	Actual cubic feet
cc/min	Cubic centimeters per minute
CFH	Standard cubic feet per hour (SCFH)
D	Diameter
FNPT	Female National Pipe Thread (U.S.)
FPM	Feet per minute
FPS	Feet per second
FS	Full scale
ft	Feet
GALS	Gallons
GPH	Gallons per hour
GPM	Gallons per minute

Table B-3 Engineering unit abbreviations (Continued)

Abbreviation	Meaning
ID	Inside diameter
I/O	Input and output systems
lbs	Pounds (English)
lbs/in^2	Pounds per square inch
LPM or L/min	Liters per minute
mL/min	Milliliters per minute
MNPT	Male National Pipe Thread (U.S.)
m/sec	Meters per second
NPT	National Pipe Thread (U.S.)
OD	Outside diameter
PSIA	Pounds per square inch absolute
PSIG	Pounds per square inch gage
RF	Raised face (flange)
RFI	Radio frequency interference
RMS	Root mean squared
sq ft	Square feet

Table B-4 Material densities

Material	Specific Gravity
ABS Pellet	0.4 to 0.6
Acrylate resin	0.4 to 0.5
Alabaster	1.33 to 1.55
Bauxite	1.07 to 1.45
Carbon black	0.25 to 0.49
Cement	1.3 to 1.4

Table B-4 Material densities (Continued)

Material	Specific Gravity
Charcoal	0.43
Chinastone	1.15 to 1.34
Clay	0.36 to 0.9
Clinker	1.3 to 1.9
Coke	0.40 to 0.49
Common salt	1.15 to 1.34
Epoxy resin	0.63 to 0.82
Ferrite	0.5 to 2.25
Glass beads	1.50 to 1.55
Glass fiber	0.2 to 0.3
Gypsum	0.7 to 1.55
Iron oxide	0.4
Limestone	1.36 to 1.57
Melamine	0.42 to 0.79
Metal cutting chips	1.5 to 1.8
Nylon resin	0.4 to 0.6
PE pellet	0.5 to 0.6
Phenolic resin	0.34 to 0.6
Polycarbonate	0.5 to 0.6
Polyester resin	0.4 to 0.6
Polyethylene pellet	0.48 to 0.50
Polypropylene pellet	0.6
Polystyrene pellet	0.58
PVC pellet	0.5 to 0.6
Salt	0.55
Sand	1.36 to 1.57
Sawdust	0.16 to 0.20

Table B-4 Material densities (Continued)

Material	Specific Gravity
Silica sand	0.8 to 1.41
Tin	3.62 to 4.12
Tungsten carbide	7.54 to 8.95
Urea	0.47 to 0.76
Wood chips	0.30 to 0.40

Table B-5 Schedule of pipe wall thickness

Nominal		Schedule							
Size	OD	5	10	40	80	100	120	140	160
1/8	.405	.035	.049	.068	.095				
1/4	.540	.049	.065	.088	.119				
3/8	.675	.049	.065	.091	.126				
1/2	.840	.065	.083	.109	.147				.187
3/4	1.050	.065	.083	.113	.154				.218
1	1.315	.065	.109	.133	.179				.250
1 1/4	1.660	.065	.109	.140	.191				.250
1 1/2	1.900	.065	.109	.145	.200				.281
2	2.375	.065	.109	.154	.218				.343
2 1/2	2.875	.083	.120	.203	.276				.375
3	3.500	.083	.120	.216	.300				.437
3 1/2	4.000	.083	.120	.226	.318				
4	4.500	.083	.120	.273	.337		.437		.531
4 1/2	5.000								
5	5.563	.109	.134	.258	.375		.500		.625

Table B-5 Pipe schedule wall thickness (Continued)

Nominal Size	OD	Schedule 5	10	40	80	100	120	140	160
6	6.625	.109	.134	.280	.432		.562		.718
7	7.625								
8	8.625	.109	.148	.322	.500	.593	.718	.812	.906
9	9.625								
10	10.750	.134	.165	.365	.593	.718	.843	1.000	1.125
11	11.750								
12	12.750	.165	.180	.406	.687	.843	1.000	1.125	1.312

Table B-6 Plastics: Specific gravity and weight

Material	Specific Gravity	Weight per Cubic Foot
HDPE	0.95	59
PVC	1.38	86
ABS	1.06	66
Fiberglass	2.00	125
Aluminum	2.65	165
Steel	7.85	490
Ductile iron	7.10	443
Concrete	2.40	150
Fresh water	1.00	62.4

Table B-7 Screwdriver insert bits (continued)

Point Size	Body Diameter	Wood Screws	Machine Screws	Sheet Metal Screws
0		No.0	No.0	No.0
		1	1	1
1	3/16	2	2	2
		3	3	3
		4	4	4
2	1/4	5	5	5
		6	6	6
		7	8	7
		8	10	8
		9	–	10
		10	–	–
3	5/16	12	12	12
		14	1/4	14
		16	5/16	5/16
4	3/8	18	3/8	3/8
		20	7/16	–
		24	1/2	–

Table B-8 Sheet metal and aluminum gauge or thickness

Gage No.	Steel	Stainless	Aluminum
7	0.179		
8	0.164	0.172	
9	0.150	0.156	
10	0.135	0.141	
11	0.120	0.125	
12	0.105	0.109	
13	0.090	0.094	0.093
14	0.075	0.078	0.079

Table B-8 Sheet metal and aluminum gauge or thickness (Continued)

Gage No.	Steel	Stainless	Aluminum
15	0.067	0.070	0.071
16	0.060	0.063	0.064
17	0.054	0.056	0.058
18	0.048	0.050	0.052
19	0.042	0.044	0.046
20	0.036	0.038	0.040
21	0.033	0.034	0.037
22	0.030	0.031	0.034
23	0.027	0.028	0.031
24	0.024	0.025	0.028
25	0.021	0.022	0.025
26	0.018	0.019	0.022

Table B-9 Steel sheet: Gauge and weight

Gauge Number	Gauge Decimal	Pounds per Square Foot of Steel Strip
6	.1943	7.926
7	.1793	7.315
8	.1644	6.707
9	.1495	6.099
10	.1345	5.487
11	.1196	4.879
12	.1046	4.267
13	.0897	3.659
14	.0747	3.047

Table B-9 Steel sheet: Gauge and weight (Continued)

Gauge Number	Gauge Decimal	Pounds per Square Foot of Steel Strip
15	.0673	2.746
16	.0598	2.440
17	.0538	2.195
18	.0478	1.950
19	.0418	1.705
20	.0359	1.465
21	.0329	1.342
22	.0299	1.220
23	.0269	1.097
24	.0239	.975
25	.0209	.853
26	.0179	.730

Table B-10 Strip and tubing: Gauge and weight

Gauge Number	Gauge Decimal	Pounds per Square Foot of Steel Strip
6	.203	8.281
7	.180	7.343
8	.165	6.731
9	.148	6.038
10	.134	5.467
11	.120	4.895
12	.109	4.447
13	.095	3.876
14	.083	3.386
15	.072	2.937

Table B-10 Strip and tubing: Gauge and weight (Continued)

Gauge Number	Gauge Decimal	Pounds per Square Foot of Steel Strip
16	.065	2.652
17	.058	2.366
18	.049	1.999
19	.042	1.713
20	.035	1.428
21	.032	1.305
22	.028	1.142
23	.025	1.020
24	.022	.898
25	.020	.816
26	.018	.734

Table B-11 Tap and drill size chart

Tap Size	Drill Size	Drill in Decimal	% of Thread
0-80	#56	.0465	73
	3/64"	.0469	71
1-64	#54	.0550	80
	#53	.0595	59
1-72	#53	.0595	67
	1/16"	.0625	50
2-56	#51	.0670	73
	#50	.0700	62
2-64	#50	.0700	70
	#49	.0730	56

Table B-11 Tap and drill size chart (Continued)

Tap Size	Drill Size	Drill in Decimal	% of Thread
3-46	#48	.0760	77
	5/64"	.0781	70
3-56	#46	.0810	69
	#45	.0820	65
4-40	#44	.0860	73
	#43	.0890	65
4-48	#42	.0935	61
	3/32"	.0938	60
5-40	#40	.0980	75
	#38	.1015	65
5-44	#38	.1015	71
	#37	.1040	63
6-32	#37	.1040	77
	7/64"	.1094	64
6-40	#34	.1110	74
	#33	.1130	69
8-32	#29	.1360	62
	#28	.1405	51
8-36	#29	.1360	70
	9/64"	.1406	57
10-24	#27	.1440	78
	#25	.1495	69
10-32	5/32"	1563	74
	#21	.1590	68
12-24	11/64"	.1719	74
	#17	.1730	73

Table B-11 Tap and drill size chart (Continued)

Tap Size	Drill Size	Drill in Decimal	% of Thread
12-28	#16	.1770	76
	#15	.1800	70
1/4-20	#7	.2010	70
	13/64"	.2031	66
1/4-28	#3	.2130	72
	7/32"	.2188	59
5/16-18	F	.2570	71
	17/64"	.2656	59
5/16-24	H	.2660	77
	I	.2720	67
3/8-16	5/16"	.3125	71
	O	.3160	68
3/8-24	21/64"	.3281	78
	Q	.3320	71
7/16-14	T	.3580	80
	23/64"	.3594	79
7/16-20	W	.3860	71
	25/64"	.3906	65
1/2-13	27/64"	.4219	72
	7/16	.4375	58
1/2-20	29/64"	.4531	65
9/16-12	15/32"	.4688	80
	31/64"	.4844	68
9/16-18	1/2"	.5000	78
	33/64"	.5156	58

Table B-12 *Aluminum 6061 alloy: Gauge and weight*

Gauge Number	Gauge Decimal	Pounds per Square Foot
6	.1620	2.286
7	.1443	2.036
8	.1285	1.813
9	.1144	1.614
10	.1019	1.438
11	.0907	1.280
12	.0808	1.140
13	.0720	1.016
14	.0641	.905
15	.0571	.806
16	.0508	.717
17	.0453	.639
18	.0403	.569
19	.0359	.507
20	.0320	.452
21	.0285	.402
22	.0253	.357
23	.0226	.319
24	.0201	.284
25	.0179	.253
26	.0159	.224

Table B-13 General materials: Specific gravity, weights

Material	Specific Gravity	Pounds per Cubic Foot
Aluminum, solid	2.64	165.0
Bakelite, solid	1.36	85
Birch wood, yellow	0.71	44
Brass, cast	8.56	534
Bronze	8.16	509
Carbon, solid	2.15	134
Cardboard	0.69	43
Cedar, red	0.38	24
Cement, Portland	1.60	94
Cherrywood, dry	0.56	35
Clay, dry excavated	1.09	68
Clay, fire	1.36	85
Clay, compacted	1.75	109
Copper, cast	8.69	542
Cork, solid	0.24	15
Cryolite	1.60	100
Elm, dry	0.56	35
Fir, Douglas	0.53	33
Glass, window	2.58	161
Gold, pure 24Kt	19.29	1204
Graphite, flake	0.64	40
Gypsum, solid	2.79	174
Ice, solid	0.92	57.4
Iron, cast	7.21	450
Iron, wrought	7.77	485

Table B-13 General materials: Specific gravity, weights (Continued)

Material	Specific Gravity	Pounds per Cubic Foot
Lead, cast	11.35	708
Leather	0.95	59
Maple, dry	0.71	44
Marble, solid	2.56	160
Oak, live, dry	0.95	59
Oak, red	0.71	44
Paper, standard	1.20	75
Pine, white, dry	0.42	26
Pine, yellow northern, dry	0.54	34
Pine, yellow southern, dry	0.72	45
Plaster	0.85	53
Redwood, California, dry	0.45	28
Resin, synthetic, crushed	0.56	35
Rubber, manufactured	1.52	95
Sand, dry	1.60	100
Sawdust	0.27	17
Silver	10.46	653
Spruce, California, dry	0.45	28
Steel, cast	7.85	490
Tin, cast	7.36	459
Water, pure	1.00	62.4

Table B-14 Wrench and driver sizes—standard (fractional inch)

Driver	For Hex Nut
1/4"	#4
5/16"	#6
11/32"	#8
3/8"	#10
7/16"	1/4"
1/2"	5/16"

INDEX

Note: **Boldface** numbers indicate illustrations.

C

D

N–O

P

Q–R

S

T

U–V

W

X–Z